Marx Joyce
Abbott Hardy Cooper Austen
Defoe Melville Chesterton Emerson Hugo
Machiavelli Eliot Grimm
Stoker Carroll Montaigne Haggard
Wilde Christie Molière
Maupassant Byron
Garnett Engels Schiller
Goethe Fitzgerald Hawthorne Kafka
Cotton Einstein Dostoyevsky Smith Hall
Baum Kipling Doyle
Dumas Henry Willis
Leslie Flaubert Nietzsche
Stockton Turgenev Balzac
Burroughs Vatsyayana Crane
Curtis Tocqueville Verne
Homer Widger Vinci
Darwin Tolstoy Busch
Potter Thoreau Gogol
Freud Zola Whitman Twain Scott
Kant Jowett Lawrence Plato Harte
Andersen Dickens Hesse
London Descartes
Poe Aristotle Wells Voltaire Burton
Hale James Hastings Cervantes Cooke
Bunner Shakespeare
Richter Chambers Irving
Doré da Benedict
Chekhov Shaw Pushkin Alcott
Dante Wodehouse
Swift Newton

tredition

tredition was established in 2006 by Sandra Latusseck and Soenke Schulz. Based in Hamburg, Germany, tredition offers publishing solutions to authors and publishing houses, combined with worldwide distribution of printed and digital book content. tredition is uniquely positioned to enable authors and publishing houses to create books on their own terms and without conventional manufacturing risks.

For more information please visit: www.tredition.com

TREDITION CLASSICS

This book is part of the TREDITION CLASSICS series. The creators of this series are united by passion for literature and driven by the intention of making all public domain books available in printed format again - worldwide. Most TREDITION CLASSICS titles have been out of print and off the bookstore shelves for decades. At tredition we believe that a great book never goes out of style and that its value is eternal. Several mostly non-profit literature projects provide content to tredition. To support their good work, tredition donates a portion of the proceeds from each sold copy. As a reader of a TREDITION CLASSICS book, you support our mission to save many of the amazing works of world literature from oblivion. See all available books at www.tredition.com.

 Project Gutenberg

The content for this book has been graciously provided by Project Gutenberg. Project Gutenberg is a non-profit organization founded by Michael Hart in 1971 at the University of Illinois. The mission of Project Gutenberg is simple: To encourage the creation and distribution of eBooks. Project Gutenberg is the first and largest collection of public domain eBooks.

Was Man Created?

Henry A. (Henry Augustus) Mott

Imprint

This book is part of TREDITION CLASSICS

Author: Henry A. (Henry Augustus) Mott
Cover design: Buchgut, Berlin – Germany

Publisher: tredition GmbH, Hamburg - Germany
ISBN: 978-3-8472-1667-4

www.tredition.com
www.tredition.de

Copyright:
The content of this book is sourced from the public domain.

The intention of the TREDITION CLASSICS series is to make world literature in the public domain available in printed format. Literary enthusiasts and organizations, such as Project Gutenberg, worldwide have scanned and digitally edited the original texts. tredition has subsequently formatted and redesigned the content into a modern reading layout. Therefore, we cannot guarantee the exact reproduction of the original format of a particular historic edition. Please also note that no modifications have been made to the spelling, therefore it may differ from the orthography used today.

FOSSIL MAN OF MENTONE. (From Popular Science Monthly, October, 1874.)

WAS MAN CREATED?

BY

HENRY A. MOTT, Jr., E.M., Ph.D., Etc.,

Member of the American Chemical Society, Member of the Berlin Chemical Society, Member of the New York Academy of Sciences, Member of the American Association for the Advancement of Science, Member of the American Pharmaceutical Association, Fellow of the Geographical Society, Etc., Etc.

Author of the "Chemists' Manual," "Adulteration of Milk," "Artificial Butter," "Testing the Value of Rifles by Firing under Water," Etc., Etc.

NEW YORK:
GRISWOLD & COMPANY,
150 Nassau Street.
1880.

Copyright by
HENRY A. MOTT, Jr.,
1880.

Trow's
Printing and Bookbinding Co.,
205-213 East 12th St.,
NEW YORK.

Electrotyped by Smith & McDougal, 82 Beekman Street, N. Y.

PREFACE.

This work was originally written to be delivered as a lecture; but as its pages continued to multiply, it was suggested to the author by numerous friends that it ought to be published in book-form; this, at last, the author concluded to do. This work, therefore, does not claim to be an exhaustive discussion of the various departments of which it treats; but rather it has been the aim of the author to present the more interesting observations in each department in as concise a form as possible. The author has endeavored to give credit in every instance where he has taken advantage of the labors of others. This work is not intended for that class of people who are so absolutely certain of the truth of their religion and of the immortality that it teaches, that they have become unqualified to entertain or even perceive of any scientific objection; for such people may be likened unto those who, "*Seeing, they see, but will not perceive; and hearing, they hear, but will not understand.*"

This work is written for the man of culture who is seeking for truth—believing, as does the author, that all truth is God's truth, and therefore it becomes the duty of every scientific man to accept it; knowing, however, that it will surely modify the popular creeds and methods of interpretation, its final result can only be to the glory of God and to the establishment of a more exalted and purer religion. All facts are truths; it consequently follows that all scientific facts are truths—there is no half-way house—a statement is either a truth or it is not a truth, according to the *law of non-contradiction*. If, therefore, we find tabulated amongst scientific facts (or truths) a statement which is not a fact, it is not science; but all statements which are facts it naturally follows are truths, and as such must be accepted, no matter how repulsive they may at first seem to some of our poetical imaginings and pet theories. We cannot help but sympathize with the feelings which prompted President Barnard to write the following lines, still we will see he was too hasty: "Much as I love truth in the abstract," he says, "I love my hope of immortality more." *** He maintained that it is better to close one's eyes to the evidences than to be convinced of the *truth* of certain doctrines which *he regards* as subversive of the fundamentals of Christian faith. "If this (is all) is the best that science can give me, then I pray no more science. Let me live on in my simple ignorance, as my fathers lived before me; and when I shall at length be summoned to my final repose, let me still be able to fold the drapery of my couch about me, and lie down to pleasant, even though they be deceitful, dreams." [1] The limitations to the acceptance of truth that President Barnard makes is wrong; for, as Professor Winchell has said, "we think it is a higher aspiration to wish to know 'the truth and the whole truth.' At the same time, we have not the slightest apprehension that the whole truth can ever dissipate our faith in a future life." [2] Let us "Prove all things and hold fast unto that which is good," recognizing the fact that "the truth-seeker is the only God-seeker."

AUTHOR

January 25, 1880.

[Pg vii]

WAS MAN CREATED?

HAECKEL'S CHART OF MAN'S DEVELOPMENT, Arranged by HENRY A. MOTT, Jr., Ph. D.

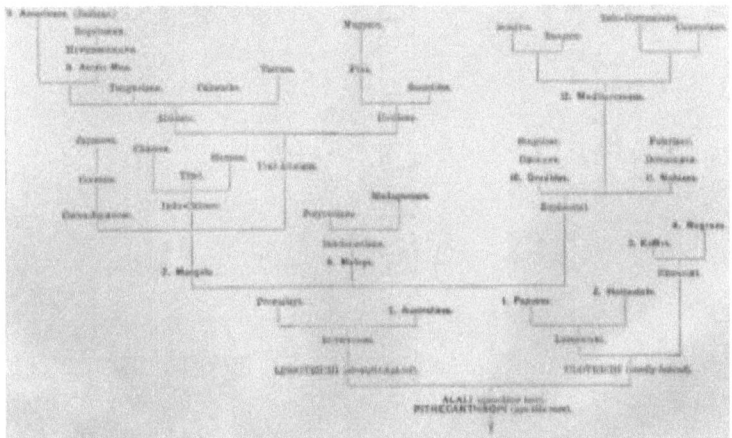

[Pg 11]

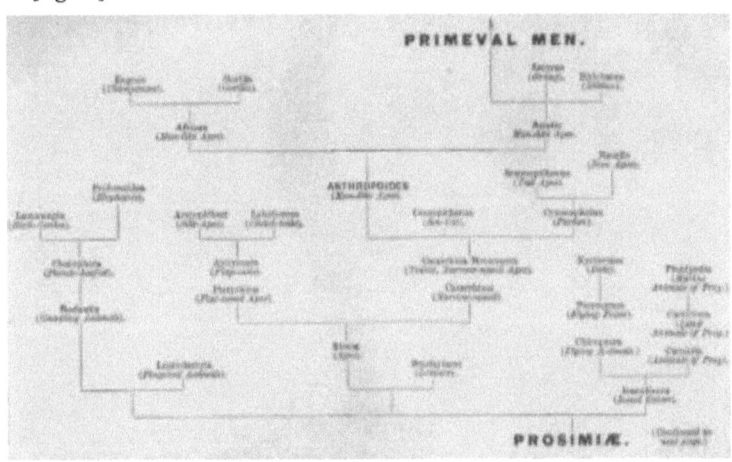

[Pg 12]

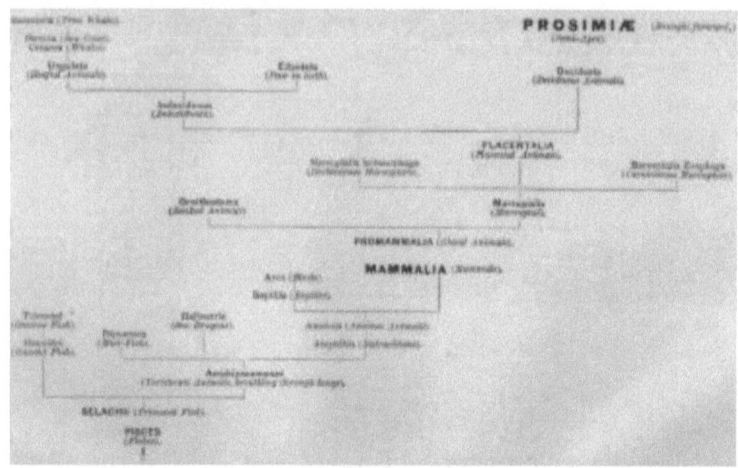

[Pg 13]

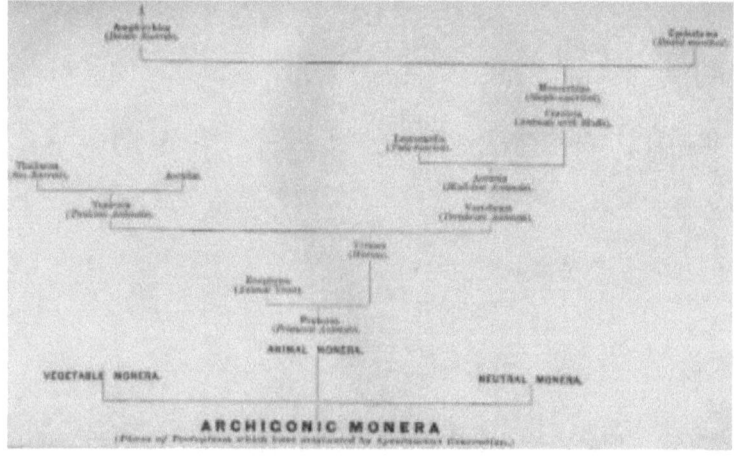

[Pg 14]

[Pg 15]

WAS MAN CREATED?

WHAT SCIENCE CAN ANSWER.

"The object of science is not to find out what we like or what we dislike—the object of science is Truth." In the discussion of the subject, *"Was Man Created?"* our object will be—not to study the many ways God might have created him, but the way he actually did create him, for all ways would be alike easy to an Omnipotent Being.

Let us look at man and ask the question: What is there about him which would need an independent act of creation any more than about the "mountain of granite or the atom of sand"? The answer comes back: Besides life, man has many mental attributes. Let us direct our attention at first to the grand phenomena of life, and then to man's attributes.

To discover the nature of life, to find out what life really is, it would be folly to commence by comparing man, the perfection of living beings, with an inorganic or inanimate substance like a brick, to discover the hidden secret; for, as Professor Orton says: [3] "That only is essential to life which is common to all forms of life. Our brains, stomach, livers, hands and feet are luxuries. They are necessary to make us human, but not living beings." Instead of man, then, it will be necessary for us to take the [Pg 16] simplest being which possesses such a phenomena; and such are the little homogeneous specks of protoplasm, constituting the Group *Monera*, which are entirely destitute of structure, and to which the name "Cytode" has been given. In the fresh waters in the neighborhood of Jena minute lumps of protoplasm were discovered by Haeckel, which, on being examined under the most powerful lens of a microscope, were seen to have no constant form, their outlines being in a state of perpetual change, caused by the protrusion from various parts of their surface of broad lobes and thick finger-like projections, which, after remaining visible for a time, would be withdrawn, to make their appearance again on some other part of the surface. To this little mass of protoplasm Haeckel has given the name *Protanæba primitiva*. These little lumps multiply by spontaneous division into two pieces, which, on becoming dependent, increase in size and acquire all the characteristics of the parent. From this illustration, it will be seen that "reproduction is a form of nutrition and a growth of the individual to a size beyond that belonging to it as an individual, so that a part is thus elevated into a (new) whole."

It is to this simple state of the monera the *fertilized* egg of any animal is transformed—the germ vesicle; the original egg kernel disappears, and the parent kernel (cytococcus) forms itself anew; and it is in this condition, a non-nucleated ball of protoplasm, a true cytod, a homogeneous, structureless body, without different constituent parts, that the human child, as well as all other living beings, take their first steps in development. No matter how wonderful this may seem, the fact stares us in the face that the entire human child, as well as every animal with all their great future possibilities, are in their first stage a small ball of this complex homogeneous substance. Whether we consider "a mere infinitesimal ovoid particle which finds space and duration enough to multiply into countless millions in the body of a [Pg 17] living fly, and then of the wealth of foliage, the luxuriance of flower and fruit which lies between this bald sketch of a plant and the gigantic pine of California, towering to the dimensions of a cathedral spire, or the Indian fig which covers acres with its profound shadow, and endures while nations and empires come and go around its vast circumference," or we look "at the other half of the world of life, picturing to ourselves the great finner whale, hugest of beasts that live or have lived, disporting his eighty or ninety feet of bone, muscle, and blubber, with easy roll, among the waves in which the stoutest ship that ever left dock-yard would founder hopelessly, and contrast him with the invisible animalcule, mere gelatinous specks, multitudes of which could in fact dance upon the point of a needle with the same ease as the angels of the schoolman could in imagination;—with these images before our minds, it would be strange if we did not ask what community of form or structure is there between the fungus and the fig-tree, the animalcule and the whale? and, *à fortiori*, between all four? Notwithstanding these apparent difficulties, a threefold unity—namely, a unity of power or faculty, a unity of form, and a unity of substantial composition—does pervade the whole living world." [4] And this unit is Protoplasm. So we see it is necessary for us to retreat to our protoplasm as a naked formless plasma, if we would find freed from all non-essential complications the agent to which has been assigned the duty of building up structure and of transforming the energy of lifeless matter into the living. Even Goethe (in 1807) almost stated this when he said: "Plants and animals, regarded in their most imperfect condition, are hardly distinguishable. This much, however, we may say, that from a condition in which plant is hardly to be distinguished from animal, creatures have appeared, gradually perfecting themselves in two [Pg 18] opposite directions—the plant is finally glorified into the tree, enduring and motionless; the animal into the human being of the highest mobility and freedom."

Let us examine for a moment this substance Protoplasm, and see in what way it differs from inorganic matter, or in what way the animate differs from the inanimate—the living from the dead.

Felix Dujardin, a French zoologist (1835) pointed out that the only living substance in the body of rhizopods and other inferior primitive animals, is identical with protoplasm. He called it *sarcode*. Hugo von Mohl (1846) first applied the name protoplasm to the peculiar serus and mobile substance in the interior of vegetable cells; and he perceived its high importance, but was very far from understanding its significance in relation to all organisms. Not, however, until Ferdinand Cohn (1850) and more fully Franz Unger (1855) had established the identity of the animate and contractile protoplasm in vegetable cells and the sarcode of the lower animals, could Max Shultz in 1856-61 elaborate the protoplasm theory of the sarcode so as to proclaim protoplasm to be the most essential and important constituent of all organic cells, and to show that the bag or husk of the cell, the cellular membrane and intercellular substance, are but secondary parts of the cell, and are frequently wanting. In a similar manner Lionel Beale (1862) gave to protoplasm, including the cellular germ, the name of "germinal matter," and to all the other substance entering into the composition of tissue, being secondary, and produced the name of "formed matter."

"Wherever there is life there is protoplasm; wherever there is protoplasm, there, too, is life." The physical consistence of protoplasm varies with the amount of water with which it is combined, from the solid form in which we find it in the dormant state to the thin watery state in which it occurs in the leaves of valisneria.

[Pg 19] As to its composition, chemistry can as yet give but scanty information; it can tell that it is composed of carbon, hydrogen, oxygen, nitrogen, sulphur, and phosphorus, and it can also tell the percentage of each element, but it cannot give more than a formula that will express it as a whole, giving no information as to the nature of the numerous albuminoid substances which compose it. Edward Cope, in his article on Comparative Anatomy, [5] gives the formula for protoplasm (as a whole), $C_{24}H_{17}N_3O_8$ + S and P, in small quantities under some circumstances. It is therefore, he says, a nitryl of cellulose: $C_{24}H_{20}O_2 + 3NH_3$. According to Mulder the composition of albumen, one of the class of protein substances to which protoplasm belongs, is $10(C_{40}H_{31}N_5O_{12}) + S_2P$. Protoplasm is identical in both the animal and vegetable kingdom; it behaves the same from whatever source it may be derived towards several re-agents, as also electricity. Is it possible, then, that the protoplasm which produces the mould is exactly the same composition as that which produces the human child? The answer is Yes, so far as the elements are concerned, but the proportions of carbon, hydrogen, etc., must enter into an infinite number of diverse stratifications and combination in the production of the various forms of life. Professor Frankland, speaking of protein, for instance, says it is capable of existing under probably at least a thousand isomeric forms. Protoplasm may be distinguished under the microscope from other members of the class to which it belongs, on account of the faculty it possesses of combining with certain coloring matters, as carmine and aniline; it is colored dark-red or yellowish-brown by iodine and nitric acid, and it is coagulated by alcohol and mineral acids as well as by heat. It possesses the quality of absorbing water in various quantities, which renders it sometimes extremely soft and nearly liquid, and sometimes hard and [Pg 20] firm like leather. Its prominent physical properties are excitability and contractility, which Kühne and others have especially investigated. The motion of protoplasm in plants was first made known by Bonaventure Corti a century ago in the Charœ plants; but this important fact was forgotten, and it had to be discovered by Treviranus in 1807. The regular motion of the protoplasm, forming a perfect current, may be seen in the hairs of the nettle, and weighty evidence exists that similar currents occur in all young vegetable cells. "If such be the case," says Huxley, "the wonderful noonday silence of a tropical forest is, after all, due only to the dullness of our hearing, and could our ears catch the murmur of these tiny maelstroms, as they whirl in innumerable myriads of living cells, which constitute each tree, we should be stunned as with a roar of a great city."

One step higher in the scale of life than the monera is the vegetable or animal cell, which arose out of the monera by the important process of segregation in their homogeneous viscid bodies, the differentiation of an inner kernel from the surrounding plasma. By this means the great progress from a simple cytod (without kernel) into a real cell (with kernel) was accomplished. Some of these cells at an early stage encased themselves by secreting a hardened membrane; they formed the first vegetable cells, while others remaining naked developed into the first aggregate of animal cells. The vegetable cell has usually two concentric coverings—cell-wall and primordial utricle. In animal cells the former is wanting, the membrane representing the utricle. As a general fact, also, animal cells are smaller than vegetable cells. Their size [6] varies greatly, but are generally invisible to the naked eye, ranging from $1/500$ to $1/10000$ of an inch in diameter. About four thousand of the smallest would be required to cover the dot put over the letter i in writing. The [Pg 21] shape of cells varies greatly; the normal form, though, is spheroidal as in the cells of fat, but they often become [7] many-sided—sometimes flattened as in the cuticle, and sometimes elongated into a simple filament as in fibrous tissue or muscular fibre.

The cell, therefore, is extremely interesting, since all animal and vegetable structure is but the multiplication of the cell as a unit, and the whole life of the plant or animal is that of the cells which compose them, and in them or by them all its vital processes are carried on. It may sound paradoxical to speak of an animal or plant being composed of millions of cells; but beyond the momentary shock of the paradox no harm is done.

The cell, then, can be regarded as the basis of our physiological idea of the elementary organism; but in the animal as well as in the plant, neither cell-wall nor nucleus is an essential constituent of the cell, inasmuch as bodies which are unquestionably the equivalents of cells—true morphological units—may be mere masses of protoplasm, devoid alike of cell-wall or nucleus. For the whole living world, then, the primary and a mental form of life is merely an individual mass of protoplasm in which no further structure is discernible. Well, then, has protoplasm been called the "universal concomitant of every phenomena of life." Life is inseparable from this substance, but is dormant unless excited by some external stimulant, such as heat, light, electricity, food, water, and oxygen.

Although we have seen that the life of the plant as well as of the animal is protoplasm, and that the protoplasm of the plant and that of the animal bear the closest resemblance, yet plants can manufacture protoplasm out of mineral compounds, whereas animals are obliged to procure it ready made, and hence in the end depend on plants. "Without plants," says Professor Orton, "animals would perish; without animals, plants had no [Pg 22] need to be." The food of a plant is a matter whose energy is all expended—is a fallen weight. But the plant organism receives it, exposes it to the sun's rays, and in a way mysterious to us converts the actual energy of the sunlight into potential energy within it. It is for this reason that life has been termed "bottled-sunshine."

The principal food of the plant consists of carbon united with oxygen to form carbonic acid, hydrogen united with oxygen to form water, and nitrogen united with hydrogen to form ammonia. These elements thus united, which in themselves are perfectly lifeless, the plant is able to convert into living protoplasm. "Plants are," says Huxley, "the accumulators of the power which animals distribute and disperse." Boussengault found long since that peas sown in pure sand, moistened with distilled water and fed by the air, obtained all the carbon necessary for their development, flowering, and fructification. Here we see a plant which not only maintains its vigor on these few substances, but grows until it has increased a millionfold or a million-millionfold the quantity of protoplasm it originally possessed, and this protoplasm exhibits the phenomena of life. This and other proof led M. Dumas to say: "From the loftiest point of view, and in connection with the physics of the globe, it would be imperative on us to say that in so far as their truly organic elements are concerned, plants and animals are the offspring of the air."

Schleiden, [8] speaking of the haymakers of Switzerland and the Tyrol, says: "He mows his definite amount of grass every year on the Alps, inaccessible to cattle, and gives not back the smallest quantity of organic substance to the soil. Whence comes the hay, if not from the atmosphere."

It has been seen, then, that plants can manufacture protoplasm, a faculty which animals are not possessed of; they at [Pg 23] best can only convert dead protoplasm into living protoplasm. Thus when vegetable or meat is cooked their protoplasm dies, but is not rendered incompetent of resuming its old functions as a matter of life. "If I," says Huxley, "should eat a piece of cooked mutton, which was once the living protoplasm of a sheep, the protoplasm, rendered dead by cooking, will be changed into living protoplasm, and thus I would transubstantiate sheep into man; and were I to return to my own place by sea and undergo shipwreck, the crustacean might and probably would return the compliment, and demonstrate our common nature by turning my protoplasm into living lobster." As has been said before, where there are life manifestations there is protoplasm. Life is regarded by one class of thinkers as the principle or cause of organization; and according to the other, life is the product or effect of organization. We must, however, agree with Professor Orton, who says: "Life is the effect of organization, not the result of it. Animals do not live because they are organized, but are organized because they are alive." In whatever way it is looked at, life is but a forced condition. "The more advanced thinkers, then, in science to-day," says Barker, "therefore look upon the life of the living form as inseparable from its substance, and believe that the former is purely phenomenal and only a manifestation of the latter. During the existence of a special force as such, they retain the term only to express the sum of the phenomena of living beings. The word life must be regarded, then, as only a generalized expression signifying the sum-total of the properties of matter possessing such organization."

In what manner, then, does this matter, possessing the phenomena of life, differ from inorganic matter, or in what manner does living matter differ from matter not living? The forces which are at work on the one side are at work on the other. The phenomena of life are all dependent upon the working of the [Pg 24] same physical and chemical forces as those which are active in the rest of the world. It may be convenient to use the terms "vitality" and "vital force" to denote the cause of certain groups of natural operations, as we employ the names of "electricity" and "electrical force" to denote others; but it ceases to do so, if such a name implies the absurd assumption that either "electricity" or "vitality" is an entity, playing the part of a sufficient cause of electrical or vital phenomena. A mass of living protoplasm is simply a machine of great complexity, the total result of the work of which, or its vital phenomena, depend on the one hand upon its construction, and on the other upon the energy supplied to it; and to speak of "vitality" as anything but the names of a series of operations is as if one should talk of the "horologity" of a clock. [9]

When hydrogen and oxygen are united by an electrical spark water is produced; certainly there is no parity between the liquid produced and the two gases. At 32° F., oxygen and hydrogen are elastic gaseous bodies, whose particles tend to fly away from one another; water at the same temperature is a strong though brittle solid. Such changes are called the properties of water. It is not assumed that a certain something called "acquosity" has entered into and taken possession of the oxide of hydrogen as soon as formed, and then guarded the particles in the facets of the crystal or amongst the leaflets of the hoar-frost. On the contrary, it is hoped molecular physics will in time explain the phenomena. "What better philosophical status," says Huxley, [10] "has vitality than acquosity. If the properties of water may be properly said to result from the nature and disposition of its molecules, I can find no intelligible ground for refusing to say that the properties of protoplasm result from the nature and disposition of its molecules."

[Pg 25] "To distinguish the living from the dead body," Herbert Spencer says, "the tree that puts out leaves when the spring brings change of temperature, the flower which opens and closes with the rising and setting of the sun, the plant that droops when the soil is dry and re-erects itself when watered, are considered alive because of these produced changes; in common with the zoophyte, which contracts on the passing of a cloud over the sun, the worm that comes to the ground when continually shaken, and the hedgehog which rolls itself up when attacked."

"Seeds of wheat produced antecedent to the Pharaohs," says Bastain, [11] "remaining in Egyptian catacombs through century after century display of course no vital manifestations, but nevertheless retain the potentiality of growing into perfect plants whenever they may be brought into contact with suitable external conditions. We must presume that either (1) during this long lapse of centuries the 'vital principle' of the plant has been imprisoned in the most dreary and impenetrable of dungeons, whither no sister effluence from the general 'soul of nature' could affect it; or else (2) that the germ of the future living plant is there only in the form of an inherited structure, whose molecular complexities are of such a kind that, after moisture has restored mobility to its atoms, its potential life may pass into actual life. Some of the lowest forms of animals and plants have such a tenacity to life that their vital manifestation may be kept in abeyance for five, ten, fifteen, or even twenty years. Though not living any more than the wheat, they also retain the potentiality of manifestation of life; and for each alike, in order that this potentiality may pass into actuality, the first requisition is water with which to restore them to that possibility of molecular rearrangement under the influence of incident forces, of which [Pg 26] the absence of water had deprived them, and without which, life in any real sense is impossible."

ANALYSIS OF A MAN.

(By Prof. Miller.)

A man 5 feet 8 inches high, weighing 154 pounds.

	lbs.	oz.	grs.
Oxygen	111	0	0
Hydrogen	14	0	0
Carbon	21	0	0
Nitrogen	3	10	0

Inorganic elements in the ash:

	lbs.	oz.	grs.
Phosphorus	1	2	88
Calcium	2	0	0
Sulphur	0	0	219
Chlorine	0	2	47

1 ounce = 437 grains.

	lbs.	oz.	grs.
Sodium	0	2	116
Iron	0	0	100
Potassium	0	0	290
Magnesium	0	0	12
Silica	0	0	2
	—	—	—
Total	154	0	0

The quantity of the substances found in a human body weighing 154 pounds:

	lbs.	oz.	grs.
Water	111	0	0
Gelatin	15	0	0
Albumen	4	3	0
Fibrine	4	4	0
Fat	12	0	0
Ashes	7	9	0
	—	—	—

| Total | 154 | 0 | 0 |

(From the "Chemists' Manual.")

[Pg 27] Professor Owen [12] says: "There are organisms (vibrieo, rotifer, macrobiotus, etc.) which we can devitalize and revitalize—devive and revive—many times. As the dried animalcule manifest no phenomena suggesting any idea contributing to form the complex one of 'life' in my mind, I regard it to be as completely lifeless as is the drowned man, whose breath and heat have gone, and whose blood has ceased to circulate. * * * The change of work consequent on drying or drowning forthwith begins to alter relations or compositions, and in time to a degree adverse to resumption of the vital form of force, a longer period being needed for this effect in the rotifer, a shorter one in the man, still shorter it may be in the amœba."

"There is," says Dumas, [13] "an eternal round in which death is quickened and life appears, but in which matter merely changes its place and form."

Let us now compare the inorganic world with the organic—the inanimate with the animate—and see if there does exist an inseparable boundary between them. The fundamental properties of every natural body are matter, form, and force. One important point to be noticed is, that the elements which compose all animate bodies are the very elements that help to build up the inanimate bodies. No new elements appear in the vegetable or animal world which are not to be found in the inorganic world. The difference between animate and inanimate bodies, therefore, is certainly not in the elements which form them, but in the molecular combination of them; and it is to be hoped that molecular physics will, at some not far distant time, enlighten us as to the peculiar state of aggregation in which the molecules exist in living [Pg 28] matter. As to the form, it is impossible to find any essential difference in the external form and inner structure between inorganic and organic bodies—for the simple monad, which is as much a living organism as the most complex being, is nothing but a homogeneous, structureless mass of protoplasm. But just as the inorganic substance, according to well-defined laws, elaborates its structure into a crystal of great beauty, so does the protoplasm elaborate itself into the most beautiful of all structures—the cell unit. Just as gold and copper crystallizes in a geometrical form, a cube—bismuth and antimony in a hexagonal, iodine and sulphur in a rhombic form—so we find among radiolaria, and among other protista and lower forms, that they "may be traced to a mathematical, fundamental form, and whose form in its whole, as well as in its parts, is bounded by definite geometrically determinable planes and angles." Now, as to the forces of the two different groups of bodies. Surely the constructive force of a crystal is due to the chemical composition, and to its material constitution. As the shape of the crystal and its size are influenced by surrounding circumstances, there is, therefore, an external constructive force at work. The only difference between the growth of an organism and that of a crystal is, that in the former case, in consequence of its semi-fluid state of aggregation, the newly added particles penetrate into the interior of the organism (inter-susception), whereas inorganic substances receive homogeneous matter from without, only by opposition or an addition of new particles to the surface. "If we, then, designate the growth and the formation of organisms as a process of life, we may with equal reason apply the same term with the developing crystal." It is for these and other reasons, demonstrating as they do the "unity of organic and inorganic nature," the essential agreement of inorganic and organic bodies in matter, form, and force, which led Tyndall [14] to say: "Abandoning [Pg 29] all disguise, the confession that I feel bound to make before you is, that I prolong the vision backward across the boundary of experimental evidence, and discern in that matter which we in our ignorance, and notwithstanding our professed reverence for its Creator, have hitherto covered with opprobrium, the promise and potency of every form and quality of life."

Returning now to our protoplasm, let us ask the question: Where did it come from? or, How did it come into existence? Though chemical synthesis has built up a number of organic substances which have been deemed the product of vitality, yet, up to the present day, the fact stands out before us that no one has ever built up one particle of living matter, however minute, from lifeless elements.

The protoplasm of to-day is simply a continuation of the protoplasm of other ages, handed down to us through periods of undefinable and indeterminable time.

The question of where protoplasm came from—how it arose—chemistry is unable to answer; but the question is answered, probably, by spontaneous generation. Only the merest particle of living protoplasm was necessary to be formed from lifeless matter in the beginning; for, in the eyes of any consistent evolutionist, any further independent formation would be sheer waste, as the hypothesis of evolution postulates the unlimited, though perhaps not, indefinite modifiability of such matter. As we have seen that there exists no absolute barrier between organic and inorganic bodies, it is not so difficult to conceive that the first particle of protoplasm may have originated, under suitable conditions, out of inorganic or lifeless matter. But the causes which have led to the origination of this particle, it may be said, we know absolutely nothing—as in the formation of the crystal and the cell—the ultimate causes remain in both cases concealed from us.

[Pg 30] At the time in the earth's history when water, in a liquid state, made its appearance on the cooled crust of the earth, the carbon probably existed as carbonic acid dispersed in the atmosphere; and from the very best of grounds, it is reasonable to assume that the density and electric condition of the atmosphere were quite different, as also the chemical and physical nature of the primeval ocean was quite different. In any case, therefore, even [15] if we do not know anything more about it, there remains the supposition, which can at least not be disputed, that at that time, under conditions quite different from those of to-day, a spontaneous generation, which is now perhaps no longer possible, may have taken place. This point is now conceded by most all of the advanced scientists of the day, and is absolutely necessary for the completion of the hypothesis of evolution.

The answer may come to this—Well, suppose the first protoplasm did originate by spontaneous generation, where did the elements or force come from which compose it?

Science has nothing to do with the coming into existence of matter or force, for she proves both to be indestructible; when they disappear, they do so only to reappear in some other form. The coming into existence of matter and force, as also the ultimate cause of all phenomena, is beyond the domain of scientific inquiry. Science has only to do with the coming in of the form of matter, not the coming in of its existence.

[Pg 31]

Fig. I.

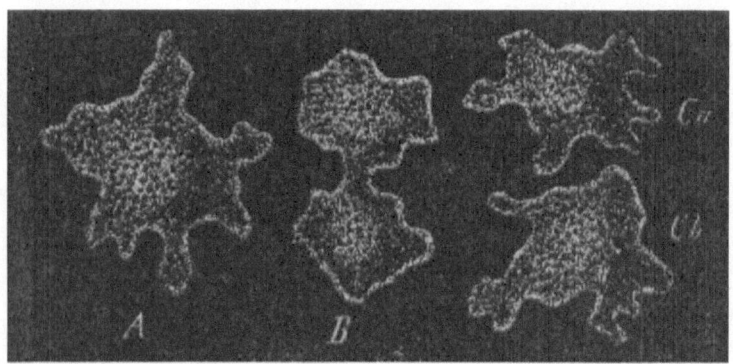

Fig. I.—A Moneron (Protamœba) in act of reproduction; *A*, the whole Moneron, which moves like ordinary Amœba, by means of variable processes: *B*, a contraction around its circumference parts it into two halves; *C*, the two halves separate, and each now forms independent individuals. (Much enlarged.) — *Haeckel.*

Fig. II.

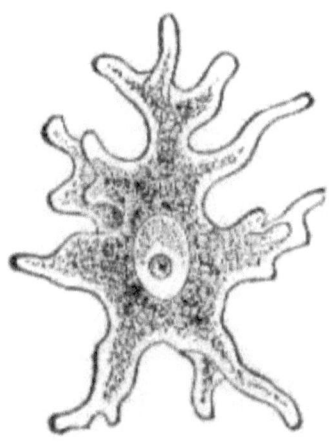 

Fig. II.—*A*, is a crawling Amœba (much enlarged).—*Haeckel*. The whole organism has the form-value of a naked cell and moves about by means of changeable processes, which are extended from the protoplasmic body and again drawn in. In the inside is the bright-colored, roundish cell-kernel or nucleus. *B*, Egg-cell of a Chalk Sponge (Olynthus).—*Haeckel*.

Fig. III.

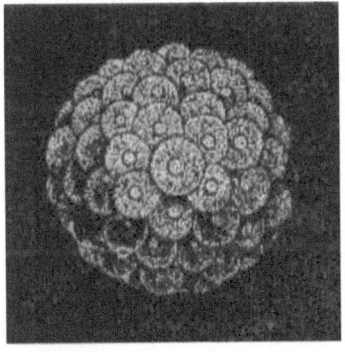

Fig. III.—Represents the next higher stage, Mulberry-germ or Morula (Synamœba).—*Haeckel*.

[Pg 32]

[Pg 33]

THE COMING INTO EXISTENCE OF MAN, BY THE SLOW PROCESS OF DEVELOPMENT.

It is necessary now to take up the little mass of living matter, admitting its coming into existence by spontaneous generation as probable, and so probable that it almost amounts to a certainty, and follow it through the many changes it is about to make under the influence of the laws which govern evolution until it has culminated in man, and these laws still acting on the brain of man, perfecting it, and leading him on to the comprehension of a grander and nobler conception of the Almighty and of his works.

The start, then, must be made with a homogeneous mass of protoplasm, such as the existing *Protamœba primitiva* of the present day, which is a structureless organism without organs, and which came into existence during the Laurentian period. It is to this simplified condition, as I have previously stated, all fertilized eggs return before they commence to develop.

The first process of adaptation effected by the monera must have been the condensation of an external crust, which, as a protecting covering, shut in the softer interior from the hostile influences of the outer world. As soon as, by condensation of the homogeneous moneron, a cell-kernel arose in the interior, and a membrane arose on the surface, all the fundamental parts of the unit were then furnished. Such a unit was an organism, [Pg 34] similar to the white corpuscle of the blood, and called *amœbæ*. Here we have two different stages of evolution; the protoplasma (better plasson) of the cytod undergoes differentiation, and is split up into two kinds of albuminous substances — the inner cell-kernel (nucleus) and the outer cell-substance (protoplasma). Edward von Benden, in his work upon *Gregarinæ*, first clearly pointed out this fact, that we must distinguish thoroughly between the plasson of cytods and the protoplasm of cells.

An irrefutable proof that such single-celled primæval animals like the amœba really existed as the direct ancestors of man, is furnished, according to the fundamental law of biogeny, by the fact that the human egg is nothing more than a simple cell.

The next step taken in advance is the division of the cell in two;—there arise from the single germinal spot two new kernel specks, and then, in like manner, out of the germinal vesicle two new cell-kernels. The same process of cell-division now repeats itself several times in succession, and the products of the division form a perfect union. This organism may be called a community of *amœbæ* (synamœbæ).

From the community of amœba morula, now arose ciliated larvæ. The cells lying on the surface extended hair-like processes or fringes of hair, which, by striking against the water, kept the whole body rotating—the lanceolate animals or amphioxus were thus first produced. Here we find from the synamœbæ which crept about slowly at the bottom of the Laurentian primeval ocean by means of movements like those of an amœba, that the newly-formed planæa by the vibrating movements of the cilia, the entire multicellular body acquired a more rapid and stronger motion, and passed over from the creeping to the swimming mode of locomotion. The planæa consisted, then, of two kinds of cells—inner ones like the amœbæ, and external "ciliated cells." The ancestors of man, which possessed the form value of the ciliated larva, is, of course, extinct at the present day.

[Pg 35]

Fig I.

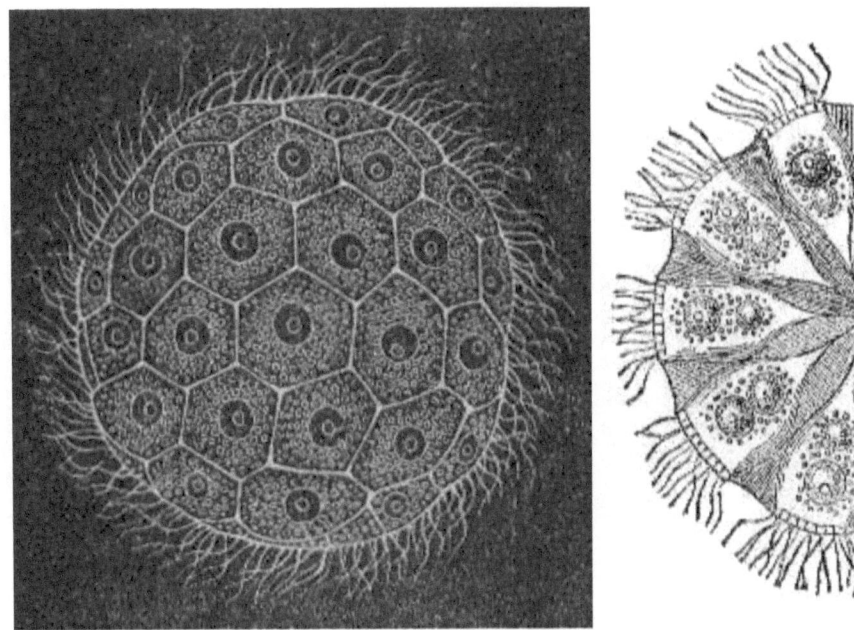

Fig. I.—The Norwegian Flimmer-ball (Magosphœra Planula), swimming by means of its vibratile fringes; seen from the surface.—*Haeckel.*

Fig. II.—The same in section. The pear-shaped cells are seen bound together in the centre of the gelatinous sphere by a thread-like process. Each cell contains both a kernel and a contractile vesicle. (Planæa Series.)—*Haeckel.*]

Fig III.   Fig. IV.

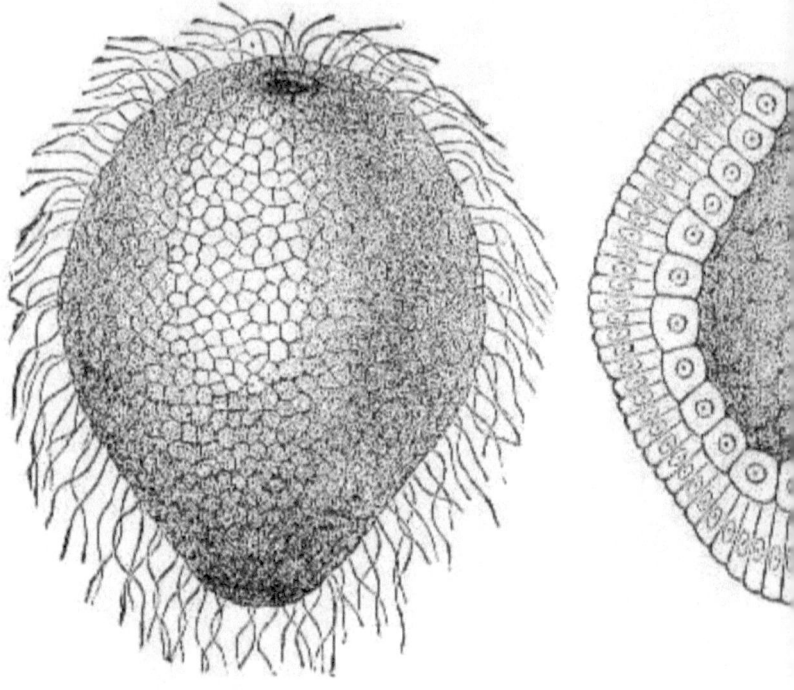

Figs. III and IV.—Represents Gastræa Series. The body consists merely of a simple primitive intestine, the wall of which is formed of two primary germ-layers.—*Haeckel*.

[Pg 36]

[Pg 37]

          Fig I.   **Fig. II.**

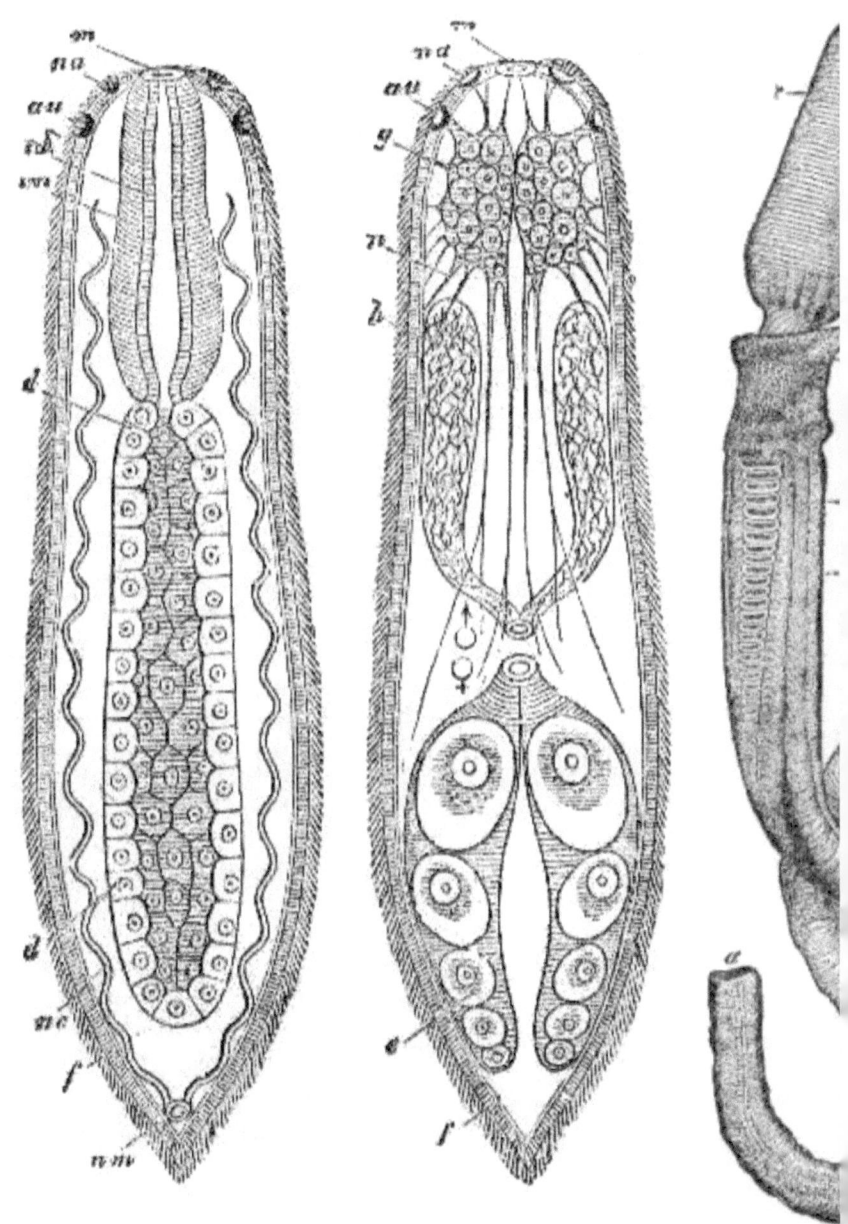

33

Figs. I and II.—Represents the next higher stage (Tubularia). Fig. I, a simple Gliding Worm (Rhabdocœlum); *m*, mouth; *sd*, throat-epithelium; *sm*, throat-muscles; *d*, stomach-intestine; *nc*, kidney-ducts; *nm*, opening of the kidneys; *au*, eye; *na*, nose-pit. Fig. II, the same Gliding Worm, showing the remaining organs; *g*, brain; *au*, eye; *na*, nose-pit; *n*, nerves; *h*, testes; ♂, male opening; ♀, female opening; *e*, ovary; *f*, ciliated outer-skin.—*Haeckel*.

Fig. III.—Represents Soft Worms (Scolecida) and is a young Acorn Worm (Balanoglossus), after *Agassiz*. *r*, acorn-like proboscis; *h*, collar; *k*, gill-openings and gill-arches of the anterior intestine, in a long row, one behind the other, on each side; *d*, digestive posterior intestine, filling the greater part of the body cavity; *v*, intestinal vessel, lying between two parallel folds of the skin; *a*, anus.

[Pg 38]

[Pg 39] Out of the planula, then, develops an exceedingly important animal form—the gastrula (that is, larva with a stomach or intestine), which resembles the planula, but differs essentially in the fact that it encloses a cavity which opens to the outside by a mouth. The wall of the progaster (primary stomach) consists of two layers of cells: an outer layer of smaller ciliated cells (outer skin or ectoderm), and of an inner layer of large non-ciliated cells (inner skin or entoderm). This exceedingly important larval form, the "gastrula," makes its appearance in the ontogenesis of all tribes of animals. These gastræada must have existed during the older primordial period, and they must have also included the ancestors of man. A certain proof of this is furnished by the amphioxus, which, in spite of its blood relationship to man, still passes through the stage of the gastrula with a simple intestine and a double intestinal wall. [16] By motion of the cilia or fringes of the skin-layer, the gastræa swam freely about in the Laurentian ocean.

The development of the gastræa now deviated in two directions—one branch of gastræads gave up free locomotion, adhered to the bottom of the sea, and thus, by adopting an adhesive mode of life, gave rise to the proascus, the common primary form of the animal plants (zoophyta). The other branch was originated by the formation of a middle germ-layer or muscular layer, and also by the further differentiation of the internal parts into various organs; more especially, the first formation of a nervous system, the simplest organs of sense, the simplest organs for secretion (kidneys), and generation (sexual organs) — this branch is the prothelmis, the common primary worms (vermes). Like the turbellaria of the present day, the whole surface of their body [Pg 40] was covered with cilia, and they possessed a simple body of an oval shape, entirely without appendages. These acœlomatous worms did not as yet possess a true body cavity (cœlom) nor blood. No member of the next higher animals are in existence, neither are there any fossil remains, owing to the soft nature of their body. They are therefore called soft worms, or scoleceda. They developed out of the turbellaria of the sixth stage by forming a true body cavity (a cœlom) and blood in their interior. The nearest still living cœlomati is probably the acorn worms (balanoglossus). The form value of this stage must, moreover, have been represented by several different intermediate stages.

Out of the four different groups of the worm tribe, the four higher tribes of the animal kingdom were developed — the star-fishes (echinoderma) and insects (arthropoda) on the one hand, and the molluscs (mollusca) and vertebrated animals (vertebrata) on the other. Out of certain cœlomati, the most ancient skull-less vertebrata were directly developed. Among the cœlomati of the present day, the ascidians are the nearest relatives of this exceedingly remarkable worm, which connect the widely differing classes of invertebrate and vertebrate animals. To these animals have been given the name of sack-worms (himatega). They originated out of the worms of the seventh stage by the formation of a dorsal nerve marrow (medulla tube), and by the formation of the spinal rod (chorda dorsalis) which lies below it. It is just the position of this central spinal rod or axial skeleton, between the dorsal marrow on the dorsal side and the intestinal canal on the ventral side, which is most characteristic of all vertebrate animals, including man, but also of the larvæ of the ascidia.

We now come to the second half of the series of human ancestors. The skull-less animal lancelet, which is still living, affords a faint idea of the members of this group (acrania). Since this little animal, in its earliest embryonic state, entirely agrees with the ascidia, and in its further development shows itself to be a true vertebrate animal, it forms a direct transition from the vertebrata to the invertebrata.

[Pg 41]

        Fig I.   **Fig. II.**   F

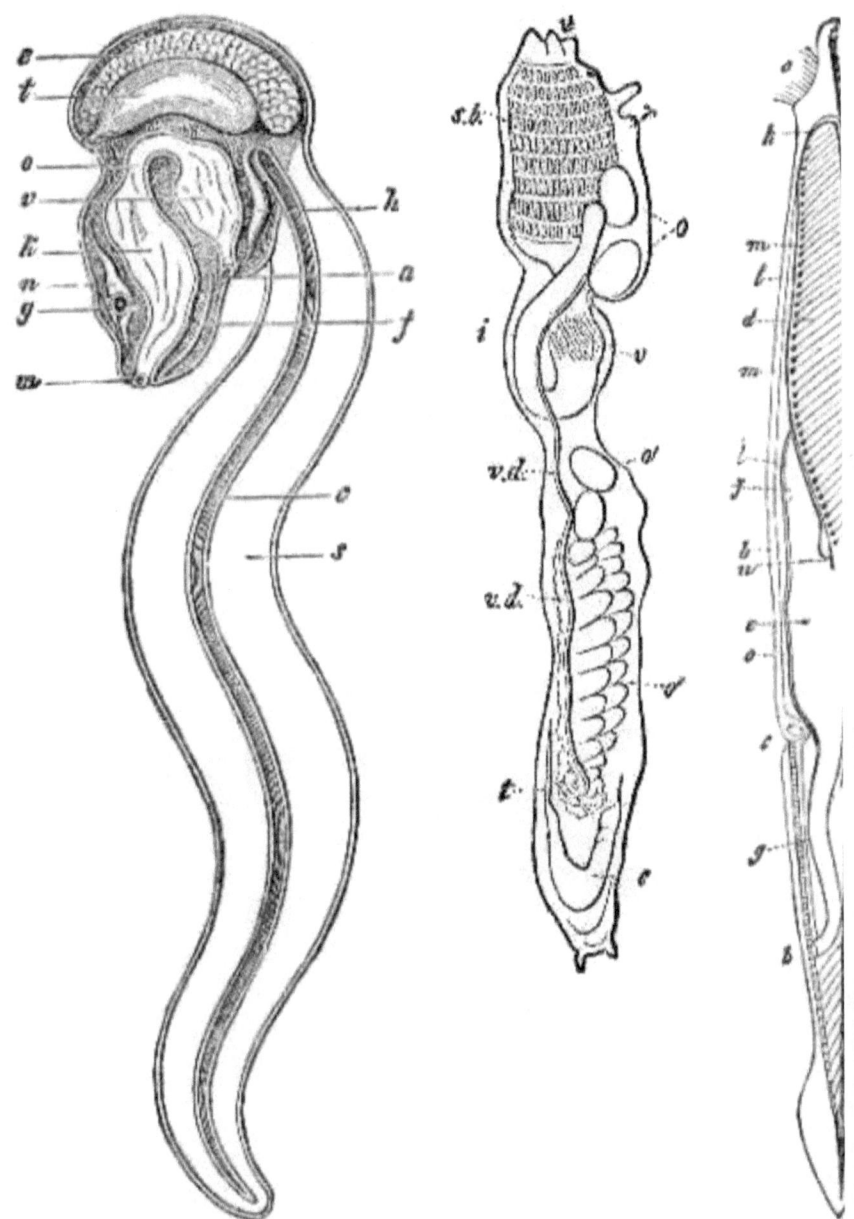

Fig. I.—Appendicularia, seen from the left side, *m*, mouth; *k*, gill intestine; *o*, œsophagus; *v*, stomach; *a*, anus; *n*, nerve ganglia (upper throat-knots); *g*, ear vesicle; *f*, ciliated groove under the gill; *h*, heart; *e*, ovary; *c*, notochord; *s*, tail.—*Haeckel*.

Fig. II.—Represents Sack Worms (Himatega), and is the structure of an Ascidian, seen from the left. *sb*, gill-sac; *v*, stomach; *i*, large intestine; *c*, heart; *t*, testes; *vd*, seed duct; *o*, ovary; *o'*, matured eggs in the body cavity. After *Milne-Edwards*.

Fig. III.—Represents the Acrania Series. Lancelet (Amhioxus Lanceolatus), twice the actual size, seen from the left. *a*, mouth-opening, surrounded by cilia; *b*, anal-opening; *c*, ventral-opening (Porus abdominalis); *d*, gill-body; *e*, stomach; *f*, liver-cœcum; *g*, large intestine; *h*, cœlum; *i*, notochord (under it the aorta); *k*, arches of the aorta; *l*, main gill-artery; *m*, swellings on its branches; *n*, hollow vein; *o*, intestinal vein.—*Haeckel*.]

[Pg 42]

[Pg 43]

Fig. I.

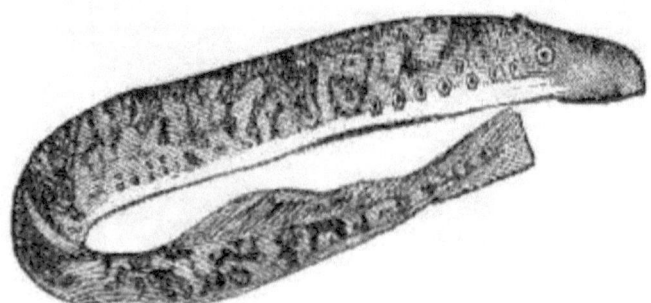

Fig. I.—Represents the Monorhina Series. Lamprey (Petromyzon Americanus) from the Atlantic—*Orton*.

Fig. II.

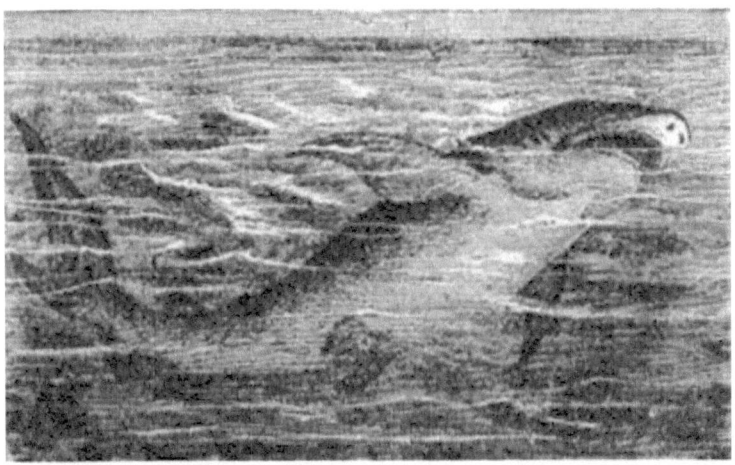

Fig. II.—Represents the Selachii. Shark (Carcharias vulgaris) from the Atlantic—*Orton*.

Fig. III.

Fig. III.—Represents the Mud-fish (Dipneusta). Lepidosiren annecteus, one-fourth natural size; African rivers.—*Orton*. Form a link between typical fishes and the Amphibians.

[Pg 44]

[Pg 45] At this stage, most probably, the separation of the two sexes began. The simpler and most ancient form of sexual propagation is through double-sexed individuals (hermaphroditismus). It occurs in the great majority of plants, but only in a minority of animals; for example, in the garden-snails, leeches, earth-worms and many other worms. Every single individual among hermaphrodites produces within itself materials of both sexes—egg and sperm. In most of the higher plants every blossom contains both the male organs (stamen and anther) and the female organs (style and germ). Every garden-snail produces in one part of its sexual gland eggs, and in another sperm. Many hermaphrodites can fructify themselves; in others, however, copulation and reciprocal fructification of both hermaphrodites are necessary for causing the development of the eggs. This latter case is evidently a transition to sexual separation (gonœhorismus).

Out of the members of the last group arose animals with skulls or craniata, having round mouths, and which are divided into hags and lampreys. The hags (myxinoides) have long cylindrical worm-like bodies. The lampreys (petromyxontes) includes those well known "nine eyes" common at the seaside.

These single-nostril animals (monorrhina) arose during the primordial period out of the skull-less animals by the anterior end of the dorsal marrow developing into the brain, and the anterior end of the dorsal skull into the skull. By the division of the single nostril of the members of the last group into two lateral halves, by the formation of a sympathetic nervous system, a jaw skeleton, a swimming bladder and two pairs of legs (breast fins or [Pg 46] fore-legs, and ventral fins or hind-legs), arose the primæval fish (selachii), which is best represented by the still-living shark (squalacei).

Out of the primæval fish arose the mud-fish (dipneusta), which is very imperfectly represented by the still-living salamander fish; the primæval fish adapting itself to land, and by the transforming of the swimming bladder into an air-breathing lung, and of the nasal cavity (which was now open into the mouth cavity) into air-passages. Their organization *might*, in some respect, be like the ceratodus and proloptems; but this is not certain.

The dipneusta is an intermediate stage between the selachii and amphibia. Out of the dipneusta arose the class of amphibia, having five toes (the pentadactyla). The gill amphibians are man's most ancient ancestors of the class amphibia. Besides possessing lungs as well as the mud-fish, they retain throughout life regular gills like the still-living proteus and axolotl. Most gilled batrachia live in North America. The paddle-fins of the dipneusta changed into five-toed legs, which were afterwards transmitted to the higher vertebrata up to man.

The gilled amphibia (sozobrachia) of the last group finally lost their gills but retained their tail, and tailed amphibians (sozura) were produced, such as the salamander and newt of the present day. Out of the sozura originated the primæval amniota (protamnia) by the complete loss of the gills by the formation of the amnion of the cochlea, and of the round window in the auditory organ, and of the organ of tears. Out of the protamnia originated the primary mammals (promammalia). The most closely related were the ornithostoma; they differed through having teeth in their jaws.

No fossil remains of the primary mammals have as yet been found, although they lived during the trias period—they possessed a very highly developed jaw. From the primary mammal arose the pouched animals (marsupialia). Numerous representatives of this group still exist: kangaroos, pouched rats and pouched dogs. The marsupial animals developed, very probably, in the mesolithic epoch (during the Jura) out of the cloacal animals; by the division of the cloaca into the rectum and the urogenital sinus, by the formation of a nipple on the mammary gland, and the partial suppression of the clavicles.

[Pg 47]

Fig. I.   Fig. II.

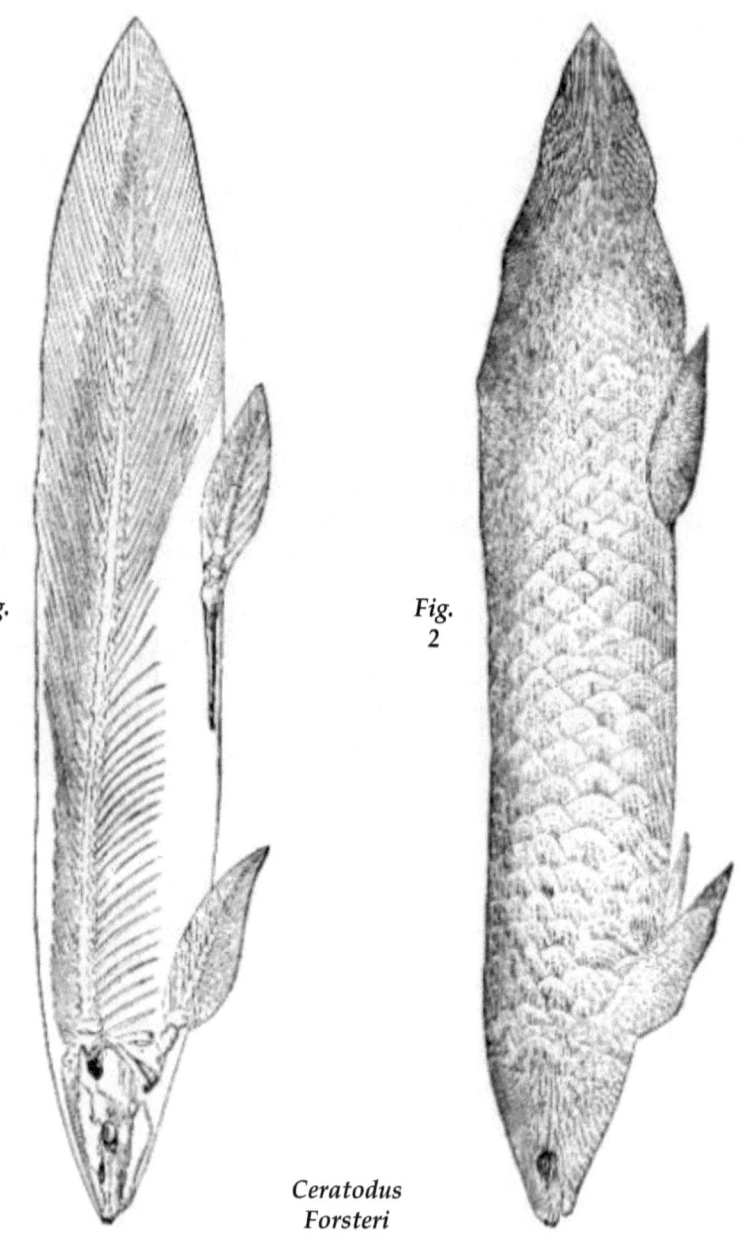

Fig. 1

Fig. 2

Ceratodus Forsteri

Figs. I and II.—The Ceratodus Forsteri occur in the swamps of Southern Australia. Form transition between fishes and Amphibia.—*Haeckel*.

[Pg 48]

[Pg 49]

Fig. I.

Fig. I.—Represents the Gilled Amphibians (Soyobranchia). The Axolotl (Siredon pisciforme), after Tegetmeier. The ordinary form with persistent branchiæ.

Fig. II.

Fig. II.—Proteus Anguinus. Europe.—*Orton*.

Fig. III.

Fig. III.—Represents the Tailed Amphibians (Soyura). Great Water-Newt (Triton cristatus), after *Bell*.

[Pg 50]

[Pg 51] From the marsupialia originated a most interesting small group of semi-apes (prosimiæ), for they are the primary forms of genuine apes and consequently of man. They developed out of handed or ape-footed marsupials (pedumana), of rat-like appearance, by the formation of a placenta, the loss of the marsupium and the marsupial bones, and by the higher development of the commissures of the brain. The still-living short-footed semi-ape (brachytarsi), especially the muki, indie and lori, possess possibly a faint resemblance.

Out of the semi-apes developed two classes of genuine apes; but as the narrow-nosed or catarrhini class are the only ones related to man, the others will not be considered. These narrow-nosed apes originated by the transformation of the jaw, and by the claws on the toes changing into nails. The still-living long-tail nose-apes and holy apes (semnopithecus) probably resembled the oldest ancestors of this group.

The tailed apes by the loss of their tail and some of their hair covering, and by the excessive development of that portion of their brain above the facial portion of the skull, developed into the man-like apes (anthropoides) — such as the gorilla and chimpanzee of Africa, and the orang and gibbon of Asia. The human ancestors of this group existed during the miocene period. From the anthropoides developed the ape-like men (pithecanthropi) during the tertiary period. The speechless primæval [Pg 52] men (alali), then, is the connecting link between the man-like apes and man. The forehand of the anthropoides became the human hand, their hinderhand a foot for walking. They did not possess the articulate human language of words and the higher developments, as consciousness and the formation of ideas must have been very imperfect.

Out of the pithecanthropi men developed genuine man, by the development of the animal language of sounds into a connected or articulate language of words — the brain also developed higher and higher. This transition took place, probably, at the beginning of the quaternary period, or possibly in the tertiary.

We have now very briefly reviewed the principal outlines of the ancestors of man, showing that man has developed from the little mass of protoplasm, as have all animals and plants. He therefore was not *spontaneously* created, but was developed. The question is often asked by simple-minded people, with much delight, Why do we not behold the interesting spectacle of the transformation of a chimpanzee into a man, or conversely of a man by retrogression into an orang?—it only shows that they are not acquainted with the first principles of the Doctrine of Descent. "Not one of the apes," says Schmidt, "can revert to the state of his primordial ancestors, except by retrogression—by which a primordial condition is by no means attained—he cannot divest himself of his acquired characters fixed by heredity, nor can he exceed himself and become man; for man does not stand in the direct line of development from the ape. The development of the anthropoid apes has taken a lateral course from the nearest human progenitors, and man can as little be transformed into a gorilla as a squirrel can be changed into a rat."

[Pg 53]

Fig. I.

Fig. I.—Salamandra Maculata.—*Haeckel*. The Water Newts and Salamanders were the next higher stage after the Proteus and the Axolotl.

[Pg 54]

Fig. I.—Represents Primæval Amniota (Protamnia). Lizard (Lacerta), after *Orton*.

Fig. II.—Represents Primary Mammals (Promammalia). Amniota Series. Duck-billed Platypus (Ornithorhynchus paradoxus).—*Haeckel*.

[Pg 56]

[Pg 57] "Feeling evidently," [17] says Haeckel, "rather than understanding, induces most people to combat the theory of their 'descent from apes.' It is simply because the organism of the ape appears a caricature of man, a distorted likeness of ourselves in a not very attractive form; because the customary æsthetic ideas and self-glorification of man are touched by this in so sensitive a point, that most men shrink from recognizing their descent from apes. It seems much pleasanter to be descended from a more highly developed divine being, and hence, as is well known, human vanity has from the earliest times flattered itself by assuming the original descent of the race from gods or demi-gods."

[Pg 58]

EVOLUTION.

In the last chapter a description was given of the various stages in man's development, from the microscopic monad up. It will be necessary now to describe briefly the various laws which have governed this evolutionary chain from the monad to man. But before proceeding directly to the subject, let us look at the doctrine of evolution as a whole, and trace it first in the formation of the world.

The doctrine of evolution is also called the theory of development—it must not, however, be confused with Darwinism—for they are not exactly synonymous. Darwinism is an attempt to explain the laws or manner of evolution. Strictly speaking, only the theory of selection should be called Darwinism, which was established in 1859. The theory of descent, or transmutation theory, or doctrine of filiation, should properly be called Lamarckism, who for the first time worked out the theory of descent as an independent scientific theory of the first order, and as the philosophical foundation of the whole science of biology.

"According to the theory of development (evolution) in its simplest form," says Henry Hartshorne, [18] "the universe as it now exists is a result of 'an immense series of changes,' related to and dependent upon each other as successive steps, or rather growths, constituting a progress; analogous to the unfolding or evolving of the parts of a growing organism." Herbert Spencer defined evolution as consisting in a progress from the homogeneous to the heterogeneous, from general to special, from the simple to the complex; and this process is considered to be traceable in the formation of worlds in space, in the multiplication of the types and species of plants and animals on the globe, in the origination and diversity of languages, literature, arts and sciences, and in all changes of human institutions and society.

[Pg 59]

Fig. I.

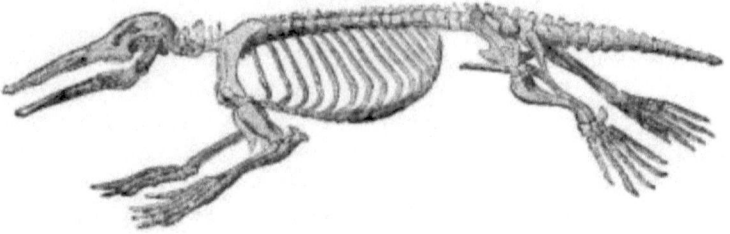

Fig. I.—Skeleton of Platypus.—*Haeckel*.

Fig. I.

Fig. I.—Represents Pouched Animals (Marsupialia). Kangaroo. (Popular Science Monthly, Feb., 1876.)

[Pg 62]

[Pg 63] Let us now apply this theory of evolution to the physical world. No determined opposition by the mass of people is likely to be manifested to the doctrine of evolution as applied to the physical world, or even to the vegetable or animal world up to man; but the minute man is included—then is a voice raised up against it, and it was for this reason that Darwin in his first work on the "Theory of Descent" did not mention man as being included in the evolutionary series. He knew too well the foolish human weakness that existed.

In a recent work by Prof. Challes, he states that he regards the material universe as "a vast and wonderful mechanism of which the least wonderful thing is its being so constructed that we can understand it."

The following is a brief description of the various theories of the world's formation:

First Theory.—By the first theory the world is supposed to have existed from eternity under its actual form. Aristotle embraced this doctrine, and conceived the universe to be the eternal effect of an eternal cause; maintaining that not only the heavens and the earth, but all animate and inanimate beings, are without beginning. To use Huxley's illustration: If you can imagine a spectator on the earth, however far back in time, he would have seen a world "essentially similar, though not perhaps in all its details, to that which now exists. The animals [Pg 64] which existed would be the ancestors of those which now exist, and like them; the plants in like manner would be such as we have now, and like them; and the supposition is that, at however distant a period of time you place your observer, he would still find mountains, lands, and waters, with animal and vegetable products flourishing upon them and sporting in them just as he finds now." This theory being perfectly inconsistent with facts, had to be abandoned.

Second Theory.—The second theory considers the universe eternal, but not its form. This was the system of Epicurus and most of the ancient philosophers and poets, who imagined the world either to be produced by fortuitous concourse of atoms existing from all eternity, or to have sprung out of the chaotic form which preceded its present state.

Third Theory.—By this theory the matter and form of the earth is ascribed to the direct agency of a spiritual cause. It is needless to say that this last theory has for its basis the popular account, generally credited to Moses in the first chapter of Genesis. I say popular, for it certainly is not a scientific account, nor was it the intention of the writer to make it so. The supposed object was to show the relation between the Creator and his works. If it had been an ultimate scientific account, the ablest minds of to-day would be unable to comprehend it, as science is progressive and constantly changing; in fifty thousand years to come, it would still appear utterly absurd. It cannot be said for this fact that the account is any the less true because it is not presented in scientific phraseology; for instance, when we remark in popular language "the sun rises," who shall say that though the expression is not astronomically true, we do not, for all practical purposes, utter as important a truth, as when we say, "The earth by its revolution brings us to that point where the sun becomes visible?" The language, also, in which the writer wrote was very imperfect; it had no equivalent to our word "air" or "atmosphere," properly speaking, for they knew not the words. "Their nearest approaches," according to J. Pye Smith, "were with words that denoted watery vapor condensed, and thus rendered visible, whether floating around them or seen in the breathing of animals; and words for smoke from substances burning; and for air in motion, wind, a zephyr whisper or a storm." It must also be remembered, "that the Hebrews had no term for the abstract ideas which we express by 'fluid' or 'matter.' If the writer had designed to express the idea, 'In the beginning God created *matter*,' he could not have found words to serve his purpose" (Phin).

[Pg 65]

Fig. I.

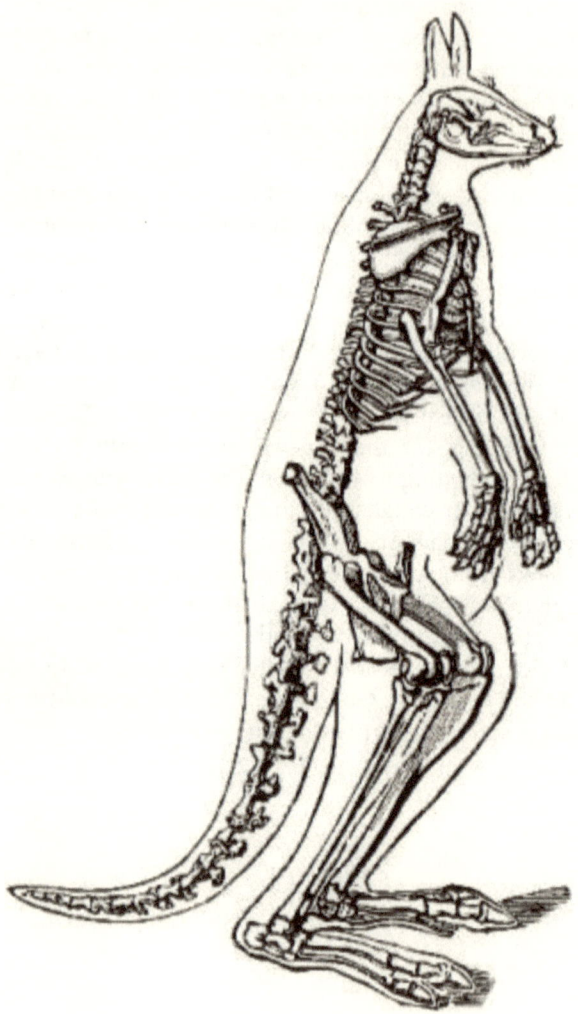

Fig. I.—Skeleton of Kangaroo. (Popular Science Monthly.)
[Pg 66]

[Pg 67]

Fig. I.

Fig. I.—Represents Semi-Apes (Prosimiæ). The Slow Loris, after *Tickel* and *Alp. Miln-Edwards*. (Natural History, by *Duncan*.)

[Pg 68]

[Pg 69] It is unnecessary to state how the Bible, which contains the so-called Mosaic account, is regarded by the different church denominations, as undoubtedly that is familiar to every one. But with respect to the view entertained by the scientist and critical school of Biblical scholars, represented chiefly by modern Germans, I may state briefly: "They regard the Bible as the human record of a divine revelation; not absolutely infallible, since there is no book written in any human language but must partake in a measure of the imperfections of that language. Many of this school, while admitting the Bible to contain the record of a true supernatural revelation, do not consider it to be without positive error of historical fact, not without false coloring from popular legend and tradition, but nevertheless a record as good as human hands could make a truly divine revelation." [19]

There is, though, a class of thinkers that altogether reject the Bible; that is to say, refuse to believe it to be a divine revelation. Hume, whom Huxley calls "the most acute thinker of the eighteenth century," thus ends one of his essays: "If we take in hand any volume of divinity or school metaphysics, for [Pg 70] instance, let us ask, *Does it contain any abstract reasoning concerning quantity or number?* No. *Does it contain any experimental reasoning concerning matter of fact and existence?* No. Commit it, then, to the flames, for it can contain nothing but sophistry and illusion." To this Huxley says: "Permit me to enforce this wise advice, Why trouble ourselves about matters of which, however important they may be, we do know nothing, and can know nothing? We live in a world which is full of misery and ignorance, and the plain duty of each and all of us is to try to make the little corner he can influence somewhat less miserable and somewhat less ignorant than it was before he entered it. To do this effectually, it is necessary to be fully possessed of only two beliefs: the first, that the order of nature is ascertainable by our faculties to an extent which is practically unlimited; the second, that our volitions count for something as a condition of the course of events. Each of these beliefs can be verified experimentally, as often as we like to try. Each, therefore, stands upon the strongest foundation upon which any belief can rest, and forms one of our highest truths."

The first words in the Mosaic account are: [20] "In the beginning God created the heaven and the earth." [21] It is seen, then, that the so-called revelation points to a beginning. The beginning referred to is an absolute beginning, for we find: "In the beginning was the Word, and the Word was with God, and the Word was God." [22] * * * "All things were made by Him; and without Him was not anything made that was made." [23] Science points also to a beginning.

Geology points to a time when man did not inhabit the earth; when for him there was a beginning. So, too, for lower organisms; so, too, for the rocky minerals; so, too, for the round [Pg 71] world itself. But the beginning that science points to is not an absolute beginning. Science has to start from some point, and that point must have a scientific foundation—the foundation of science is matter, which is inseparable from form and force. Natural science teaches that matter is eternal and imperishable; for experience has never shown us that even the smallest particle of matter has come into existence or passed away. "A naturalist," says Haeckel, "can no more imagine the coming into existence of matter than he can imagine its disappearance, and he therefore looks upon the existing quantity of matter in the universe as a given fact." "The creation of matter, if, indeed," says Haeckel, [24] "it ever took place, is completely beyond human comprehension, and can therefore never become a subject of scientific inquiry. We can as little imagine a *first beginning* of the eternal phenomena of the motion of the universe as of its final end." [25] It is evident, then, that the absolute beginning of the universe and its absolute end are not questions of science, and can be known only as revealed by faith. Paul says: "By faith we understand that the world was framed by the word of God, so that things which are seen were not made of things which appeared." [26]

[Pg 72]

Fig. I.

Fig. I.—Represents Tailed Apes (Menocerca). Proboscis Monkey (Presbytes larvatus). (Mammalia.)—*Louis Figuier.*

The natives of Borneo pretend that these monkeys, or, as sometimes called, Kahan, are men who have retired to the woods to avoid paying taxes; and they entertain the greatest respect for a being who has found such ready means of evading the responsibilities of society.—*Figuier.*

[Pg 73]

[Pg 74]

Fig. I.

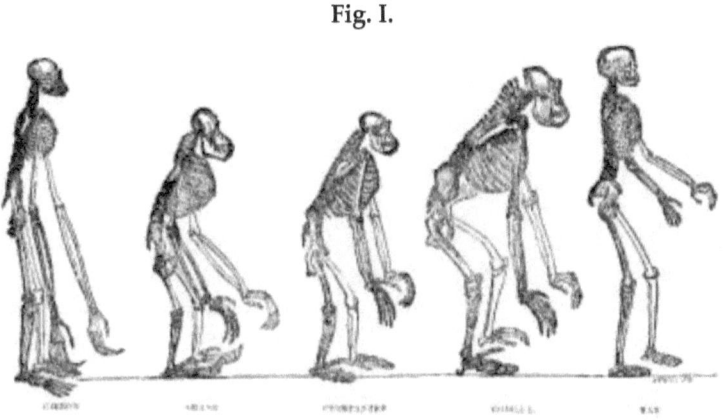

Fig. I.—Photographically reduced from diagrams of the natural size (except that of the Gibbon, which was twice as large as nature), drawn by *Waterhouse Hawkins*, from specimens in the museum of the Royal College of Surgeons. (*Huxley's* "Man's Place in Nature.")

[Pg 75]

If, therefore, science makes the "history of creation" its highest and most difficult and most comprehensible problem, it must deal with "*the coming into being of the form* of natural bodies." Let us look for a minute at Kant's Cosmogony, or, as Haeckel says, [27] Kant's Cosmological Gas Theory: "This wonderful theory," says Haeckel, "harmonizes with all the general series of phenomena at present known to us, and stands in no irreconcilable contradiction to any one of them. Moreover, it is purely mechanical and monistic, makes use exclusively of the in [Pg 76] herent forces of eternal matter, and entirely excludes every supernatural process, every prearranged and conscious action of a personal creator." Compare this last statement with the following: "I will, however," says Haeckel, [28] "not deny that Kant's grand cosmogony has some weak points." * * * "A great unsolved difficulty lies in the fact that the cosmological gas theory furnishes no starting-point at all in explanation of the first impulse which caused the rotary motion in the gas-filled universe."

Whewell [29] has pointed out, that the nebular hypothesis is null without a creative act to produce the inequality of distribution of cosmic matter in space.

It is seen, then, that according to Kant's theory we are to suppose that millions of years ago there appeared a nebulous mass possessing a rotary motion, and unequally distributed through space. This is what science calls a beginning, and may assert that every physical event of a hundred million of ages existed potentially in that nebulous mass. But this is really no explanation of the ultimate and real cause of anything. Reason demands the cause of this beginning, the source that gave to the nebulous mass its rotary motion; the power that distributed the matter in space; the antecedents of the cosmical vapor. In absence of antecedents, what was the cause of this fire-mist — of these forces active in it? Reason will never remain satisfied until these questions are answered. But physical science can trace the thread no further back, and must be dumb to all ulterior inquiries. It is true, then, as physicists assert, "that their science does not mount actually to God."

[Pg 77]

Fig. I.

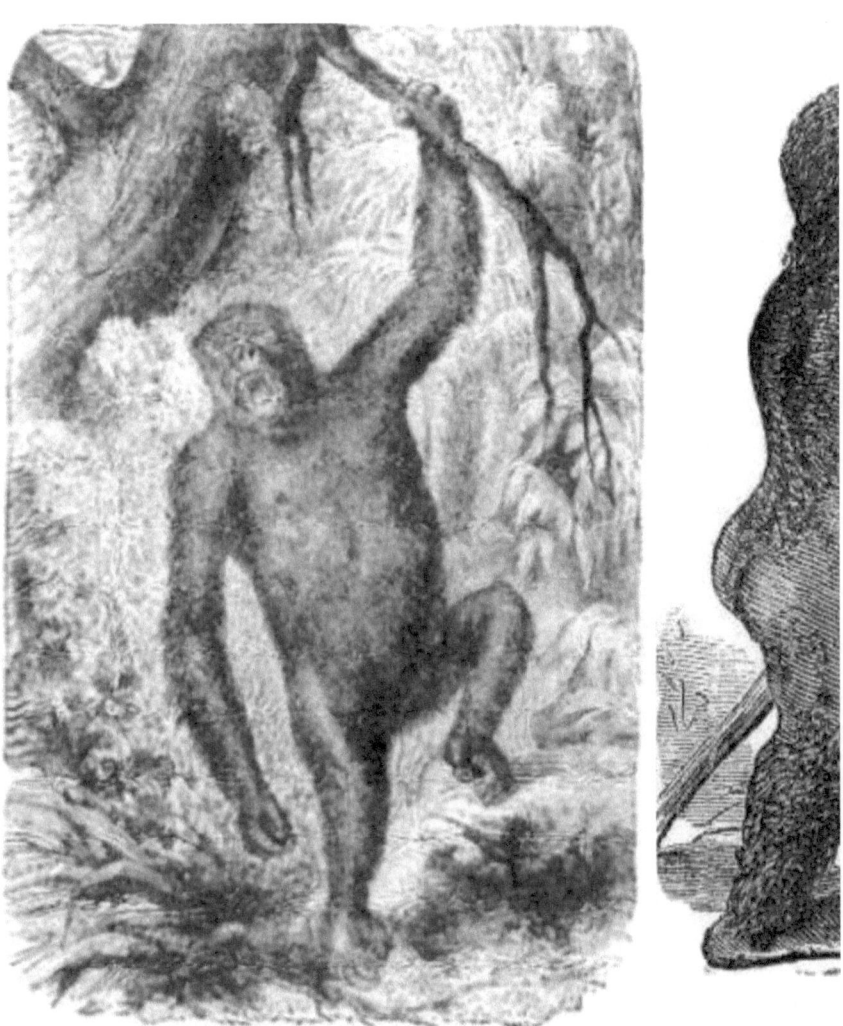

Fig. I. — Represents Man-like Apes (Anthropoides). The Male Gorilla. (Natural History, by *Duncan*.)

Fig. II. — Represe

Imaginative. (

Fig. I.

Fig. I.—The Monkey Men of Dourga Strait. (Natural History, by *Rev. Dr. Wood.*)

To God then, in strict accordance with our reason, is to be attributed not only the origination of matter, but all its future developments. When I speak of matter, it must be understood [Pg 81] that I mean force; for "if matter were not force, and immediately known as force, it could not be known at all, could not be rationally inferred. The operation of force could furnish no evidence of the existence of forceless matter. If force is not matter, then force can exist and operate without matter; its existence and operation are no evidence of the existence of matter. And as matter is forceless, it can itself give no evidence of its own existence, for that would be an exercise of force. If force cannot exist and operate without matter, then force depends for its existence and operation on the forceless, which destroys itself; or force depends for its existence on matter as some property or force, and so matter and force are identified, and force depends on itself only, as it must." [30] The idea, then, that force is an attribute of matter and inherent in it, is absurd, for there is not a shadow of evidence that force is or can be an attribute of matter. We have no knowledge of the origin of any force save of that which emanates from human volition. All our knowledge of force presents it as an effort of intelligent will. "We are driven," says Winchell, "by the necessary laws of thought, to pronounce those energies styled gravitation, heat, chemical affinity and their correlates, nothing less than intelligent will. But as it is not human will which energizes in whirlwind and the comet, it must be divine will." "In all cases, the creative power of God is an act of power, and the power does not perish with its inception, but continues to operate until the act is reversed and undone; so that everything that God has created constitutes a positive and intrinsic force, though borrowed from Him. Every incident runs back to God as its originator and real cause. The true philosophical doctrine makes God distinct from all his works, and yet acting in them. This doctrine has been held by the greatest thinkers [Pg 82] the world has ever produced, such as Descartes, Lerbrisky, Berkeley, Herschel, Faraday, and a multitude of others." "It seems to be required," says Dr. McCosh, "by that deep law of causation which not only prompts us to seek for a law in everything but an adequate cause, to be found only in an intelligent mind." "Our greatest American thinker, Jonathan Edwards," says Dr. McCosh, (whom I can claim as my predecessor,) "maintains that, as an image in a mirror is kept up by a constant succession of rays of light, so nature is sustained by a constant forth-putting of the divine power. In this view Nature is a perpetual creation. God is to be seen not only in creation at first, but in the continuance of all things." "They continue to this day according to Thine ordinances."

Returning now to the history of the creation given by Moses, Haeckel says, "Although Moses looks upon the results of the great laws of organic development as the direct actions of a constructing Creator, yet in his theory there lies hidden the ruling idea of a progressive development and a differentiation of the originally simple matter. We can therefore bestow our just and sincere admiration on the Jewish lawgiver's grand insight into nature, without discovering in it a so-called 'divine revelation.' That it cannot be such is clear from the fact that two great fundamental errors are asserted in it, namely, first the *geocentric* error, that the earth is the fixed central point of the whole universe, round which the sun, moon and stars move; and secondly, the *anthropocentric* error that man is the premeditated aim of the creation of the world, for whose service alone all the rest of nature is said to have been created. The former of these errors was demolished by Copernicus' System of the Universe in the beginning of the sixteenth century, the latter by Lamarck's Doctrine of Descent in the beginning of the nineteenth century."

[Pg 83]

Fig. I.

Fig. I. — Australian Savage. — *Orton*.   Fig. II.

Fig. III.

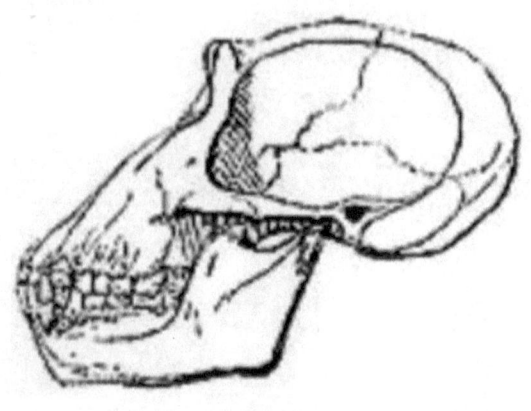

Fig. III.—Skull of Chimpanzee (Troglodytes niger).

Fig. V.

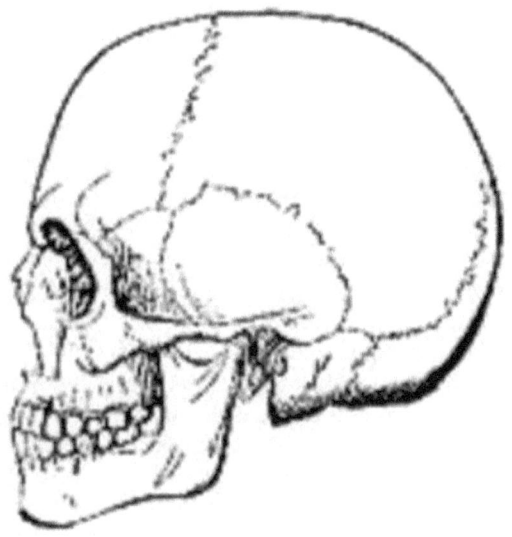

Fig. V.—Skull of European.

[Pg 84]

[Pg 85] Prof. Huxley, in his lecture on "Evidences of Evolution," spoke of the Mosaic account as Milton's hypothesis. First, "because," says Huxley, "we are now assured upon the authority of the highest critics, and even of dignitaries of the church, that there is no evidence whatever that Moses ever wrote this chapter, or knew anything about it;" and second, as this hypothesis is presented in Milton's work on "Paradise Lost," it is appropriate to call it the Miltonic Hypothesis. "In the Miltonic account," says Huxley, "the order in which animals should have made their appearance in the stratified rocks would be this: Fishes, including the great whale, and birds; after that all the varieties of terrestrial animals. Nothing could be further from the facts as we find them. As a matter of fact we know of not the slightest evidence of the existence of birds before the jurassic and perhaps the triassic formations. If there were any parallel between the Miltonic account and the circumstantial evidence, we ought to have abundant evidence in the devonian, the silurian, and carboniferous rocks. I need not tell you that this is not the case, and that not a trace of birds makes its appearance until the far later period which I have mentioned. And again, if it be true that all varieties of fishes, and the great whales and the like, made their appearance on the fifth day, then we ought to find the remains of these things in the older rocks—in those which preceded the carboniferous epoch. Fishes, it is true, we find, and numerous ones; but the great whales are absent, and the fishes are not such as now live. Not one solitary species of fish now in existence is to be found there, and hence you are introduced again to the difficulty, to the dilemma, that either the creatures that were created then, which came into existence the sixth day, were not those which are found at present, or are not the direct and immediate predecessors of those which now exist; but in that case you must either have [Pg 86] had a fresh species of which nothing has been said, or else the whole story must be given up as absolutely devoid of any circumstantial evidence."

It is for these and many other reasons that I feel bound to omit the Mosaic account, no matter how near some portions of it coincide with the facts the earth has opened out to the scientist.

KANT'S COSMOGONY.

It is maintained by Kant's Cosmogony that every substance, be it solid or liquid, constituting the entire universe, was, inconceivable ages ago, in their homogeneous gaseous or nebulous condition. Owing to an impulse being given to the nebulous mass, it acquired a rotary movement, which divided the nebulous mass up into a number of masses which, owing to the rotation, acquired greater density than the remaining gaseous mass, and then acted on the latter as central points of attraction. Our solar system was thus a gigantic gaseous or nebulous ball, all the particles of which revolved around a common central point — the solar nucleus. This nebulous ball assumed by its continual rotation a more or less flattened spheroidal form. By the continual revolution of this mass, under the influence of the centripetal and centrifugal forces, a circular nebular ring separated (like the present ring around Saturn) from the rotating ball. In time the nebulous ring condensed to a planet, which began to revolve around its own axis. When the centrifugal force became more powerful than the centripetal force in the planet, rings were formed, which, in turn, formed planets which revolved around their axes, as also around their planets, as the latter moved around the sun, and thus arose the moons, only one of which moves around our earth, while four move around Jupiter and six around Uranus. This order of things was repeated over and [Pg 87] over again until thereby arose the different solar systems — the planets rotating around their central suns, and the satellites or moons moving around their planets. By a continuous increasing of refrigeration and condensation, a fiery fluid or molten state occurred in these rotating bodies. They then emitted an enormous amount of heat by rapid condensation, and the rotating bodies — suns, planets, and moons — soon became glowing balls of fire, emitting light and heat. The $1/1000$ part of a pound of magnesium wire, burning in the open air, will give a light which will last during one second, and can be seen at a distance of thirty miles; imagine, then, what the light would be from these huge balls of fire floating through space. The earth forms a small part — nay, even the sun whose mass is equal to 354,936 earths like ours, is but an infinitesimal portion of the whole. By the continual emitting of heat, however, these fiery balls had a crust form on the outside, which enclosed a fiery fluid nucleus. The crust for a time must have been a smooth sheet, but afterward very uneven, having protuberances and cavities form over its surface, owing to the molten mass within becoming condensed and contracted; the crust not following this change sufficiently close, must have fallen in, and thus produced the cavities.

[Pg 88]

Mongolian.   Malay.

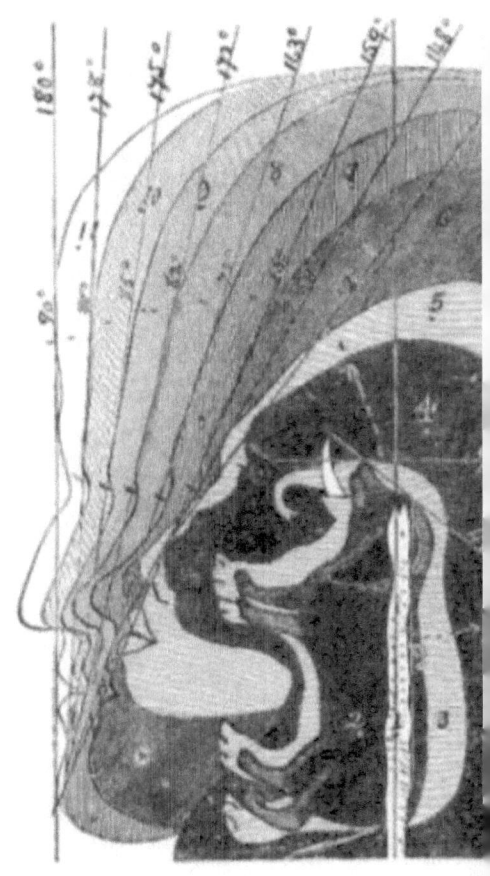

Facial Angle, by *Prof. Nelso*
1, Snake; 2, Dog; 3, Elephant
5, Human Idiot; 6, The Bushman; 7, The Uncu
9, The Civilized; 10, The Enlightened; 11, The

Caucasian
(after *Van Evrie*).

Head of Nose-Ape
(after *Brehm*).

[Pg 89]

All the time, by the condensation, the diameter of the earth was being diminished. The irregular cooling of the crust caused irregular contractions on the surface, and as the diameter of the molten mass within was continually diminishing, many elevations and depressions were caused, which were the foundations of mountains and valleys.

After the temperature of the earth had been reduced by the thickening of the crust—when it became sufficiently cool—the water which existed in steam was condensed and precipitated, falling in torrents, washing down the elevations, filling the depressions with the mud carried along, and depositing it in layers. [Pg 90] It was not until the earth became covered with water that life was possible in any form, as both animals and plants consist to a very great extent of water. At this stage in the history of the earth, then, the little mass of protoplasm, which we have spoken so much about, came into existence in all probability, as has been stated, by spontaneous generation.

LAWS OF EVOLUTION.

Let us now examine some of the laws of evolution, as also some of the connecting links which blend one stage of man's development with another, which at first thought would seem unexplainable.

Haeckel [31] summarizes the inductive evidences of Darwinism as follows: 1. Paleontological series (phylogeny); 2. Embryological development of the individual (ontogeny); 3. The correspondence in the terms of these two series; 4. Comparative anatomy (typical forms and structures); 5. Correspondence between comparative anatomy and ontogeny; 6. Rudimentary organs (dipeliology); 7. The natural system of organisms (classification); 8. Geographical distribution (chorology); 9. Adaptation to the environment (œcology); 10. The unity of biological phenomena.

It will of course be impossible to consider even hastily all of the inductive evidence belonging to the several groups mentioned above, for the scope of this work would not permit of it. Only such facts as present themselves most forcibly to the mind will be considered.

Darwinism, as has already been stated, is not the doctrine of evolution; it is, however, a successful attempt to explain the law or manner of evolution. The *law of natural selection*, pointed [Pg 91] out by Darwin, is called by Herbert Spencer, *The struggle for existence*. Darwin discovered that natural selection produces fitness between organisms and their circumstances, which explains the law of *the survival of the fittest*.

It is a well-known fact that man can, by pursuing a certain method of breeding or cultivation, improve and in various ways modify the character of the different domestic animals and plants. By always selecting the best specimen from which to propagate the race, those features which it is desired to perpetuate become more and more developed; so that what are admitted to be real varieties sometimes acquire, in the course of successive generations, a character as strikingly distinct, to all appearances, from those of the varieties, as one species is from another species of the same genus. It is evident that both natural and artificial selection depends on adaptation and inheritance. The difference between the two forms of selection is that, in the first case, the will of man makes the selection according to a plan, whereas in natural selection the struggle for life and the survival of the fittest acts without a plan other than that the most adaptable organism shall survive which is most fit to contend with the circumstances under which it is placed. Natural selection acts, therefore, much more slowly than artificial selection, although it brings about the same end. Adaptation in the struggle for life is an absolute necessity.

In every act of breeding, a certain amount of protoplasm is transferred from the parents to the child, and along with it there is transferred the individual peculiar molecular motion. Adaptation or transmutation depends upon the material influence which the organism experiences from its surroundings, or its conditions of existence; while the transmission from inheritance is due to the partial identity of producing and produced organisms.

Organized beings, as a rule, are gifted with enormous powers [Pg 92] of increase. Wild plants yield their crop of seed annually, and most wild animals bring forth their young yearly or oftener. Should this process go on unchecked, in a short time the earth would be completely overrun with living beings. It has been calculated that if a plant produces fifty seeds (which is far below the reproductive capacity of many plants) the first year, each of these seeds growing up into a plant which produces fifty seeds, or altogether two thousand five hundred seeds the next year, and so on, it would under favorable conditions of growth give rise in nine years to more plants by five hundred trillions than there are square feet of dry land upon the surface of the earth.

Slow-breeding man has been known to double his number in twenty-five years, and according to Euler, this might occur in little over twelve years. But assuming the former rate of increase, and taking the population of the United States at only thirty millions, in six hundred and eighty-five years their living progeny would have each but a square foot to stand upon, were they spread over the entire globe, land and water included. But millions of species are doing the same thing, so that the inevitable result of this strife cannot be a matter of chance. Evidently those individuals or varieties having some advantage over their competitors will stand the best chance to live, while those destitute of these advantages will be liable to destruction. Nature may be said (metaphorically) to choose (like the will of man in artificial selection) which shall be preserved and which destroyed.

That portion of the theory of development which maintains the common descent of all species of animals and plants from the simplest common origin, I have already stated with full justice should be called Lamarckism. Progress is recognized by all scientists to be a law of nature. Some of the more [Pg 93] important facts which sustain the theory of development, I propose now to present as briefly as possible.

RUDIMENTARY ORGANS.

One of the strongest arguments in favor of the hypothesis of a genetic connection among all animals (including man), at least among all those belonging to the same great types, is the presence of rudimentary parts. By rudiments in anatomy are meant organs or structures imperfectly developed, so as to be almost or entirely without functional use. "Each of them represents in germ, as it were, in one animal (or plant), that which is perfect and useful in another type."

For a few examples: The little fold of caruncle at the inner margin of the eye in man, represents the nictitating membrane of birds. Eyes which do not see form a striking example. These are found in very many animals which live in the dark, as in caves or underground. Their eyes are often perfectly developed but are covered by a membrane, so that no ray of light can enter and they can never see. Such eyes, without the function of sight, are found in several species of moles and mice which live underground, in serpents and lizards, in amphibious animals (proteus, cæcilia) and in fishes; also in numerous invertebrate animals which pass their lives in the dark, as do many beetles, crabs, snails, worms, etc.

Other rudimentary organs are the wings of animals which cannot fly. For example, the wings of the running birds, like the ostrich, emeu, cassowary, etc., the legs of which become exceedingly developed. The muscles which move the ears of animals are still present in man, but of course are of no use; by continual practice persons have been able to move their ears by these muscles. The rudiment of the tail of animals which man [Pg 94] possesses in his 3-5 tail vertebræ, is another rudimentary part—in the human embryo it stands out prominently during the first two months of its development; it afterwards becomes hidden. "The rudimentary little tail of man is irrefutable proof that he is descended from tailed ancestors." In woman the tail is generally, by one vertebra, longer than in man. There still exists rudimentary muscles in the human tail which formerly moved it.

Another case of human rudimentary organs, only belonging to the male, and which obtains in like manner in all mammals, is furnished by the mammary glands on the breast, which, as a rule, are active only in the female sex. However, cases of different mammals are known, especially of men, sheep and goats, in which the mammary glands were fully developed in the male sex, and yield milk as food for their offspring. The vermiform appendix of the large intestine in man, is another illustration of a part which has no use, but in one marsupial is three times the length of its body. The rudimentary covering of hair over certain portions of the body, is not without interest. Over the body we find but a scanty covering, which is thick only on the head, in the armpits, and on some other parts of the body. The short hairs on the greater part of the body are entirely useless, and are the last scanty remains of the hairy covering of our ape ancestors. Both on the upper and lower arm the hairs are directed toward the elbow, where they meet at an obtuse angle — this striking arrangement is only found in man and the anthropoid apes, the gorilla, chimpanzee, orang, and several species of gibbons. The fine short hairs on the body become developed into "thickset, long, and rather coarse dark hairs," when abnormally nourished near old-standing inflamed surfaces. [32] The fine wool-like hair or so-called lanugo with which the human fœtus, during the fifth and [Pg 95] sixth months, is thickly covered, offers another proof that man is descended from an animal which was born hairy, and remained so during life. This covering is first developed during the fifth month, on the eyebrows and face, and especially around the mouth, where it is much longer than that on the head. Three or four cases have been recorded of persons born with their whole bodies and faces thickly covered with fine long hairs. Prof. Alex. Brandt compared the hair from the face of a man thus characterized, aged thirty-five, with the lanugo of a fœtus, and finds it quite similar in texture. Eschricht [33] has devoted great attention to this rudimentary covering, and has thrown much light on the subject. He showed that the female as well as the male fœtus possessed this hairy covering, showing that both are descended from progenitors, both sexes of whom were hairy. Eschricht also showed, as stated above, that the hair on the face of the fifth month fœtus is longer on the face than on the head, which indicates that our semi-human progenitors were not furnished with long tresses, which must therefore have been a late acquisition. The question naturally arises, is there any explanation for the loss of hair covering?

[Pg 96]

Fig. I.—The Hairy-Faced Burmese Family. (From Scientific American, Feb. 20, 1875.)

[Pg 97]

Darwin is of the opinion that the absence of hair on the body is, to a certain extent, a secondary sexual character; for, in all parts of the world, women are less hairy than men. He says: "Therefore we may reasonably suspect that this character has been gained through sexual selection." As the body in woman is less hairy than in man, and as this character is common to all races, we may conclude that it was our female semi-human ancestors who were first divested of hair.

Professor Grant Allen [34] has given much study to the subject of the loss of hair in the human being; and his investigations [Pg 98] are worthy of careful consideration. He shows conclusively that those parts of an animal which are in constant contact with other objects are specially liable to lose their hair. This is noticeable on the under surface of the body of all animals which habitually lie on the stomach. The soles of the feet of all mammals where they touch the ground are quite hairless; the palms of the hands in the quadrumana present the same appearance. The knees of those species which frequently kneel, such as camels and other ruminants, are apt to become bare and hard-skinned. The friction of the water has been the means of removing the hair from many aquatic mammals—the whales, porpoises, dugongs, and manatees are examples.

As the back of man forms the specially hairless region of his body, we must conclude that it is in all probability the first part which became entirely denuded of hair. The gorilla, according to Professor Gervais, is the only mammal which agrees with man in having the hair thinner on the back, where it is partly rubbed off, than on the lower surface. Du Chaillu states that he has "himself come upon fresh traces of a gorilla's bed on several occasions, and could see that the male had seated himself with his back against a tree-trunk." He also says: "In both male and female the hair is found worn off the back; but this is only found in very old females. This is occasioned, I suppose, by their resting at night against trees, at whose base they sleep." The gorilla has only very partially acquired the erect position, and probably sits but little in the attitude common to man. In man the case is different; in proportion as his progenitors grew more and more erect, he must have lain less and less upon his stomach, and more and more upon his back or sides, and this is seen in the savage man during his lazy hours—who stretches himself on the ground in the sun, with his back propped, where possible, by a slight mound or the wall of his hut. The con [Pg 99] tinual friction of the surface of the back would arrest the growth of hair; for hair grows where there is normally less friction, and *vice versâ*.

As man became more and more hairless, especially among savage and naked races, we should conclude that such a modification would be considered a beauty, and women would select such men in preference to more hairy individuals. The New Zealand proverb is: "There is no woman for a hairy man." Sexual selection, then, would play a very important part; and the difficulty of understanding how man became divested of hair is readily explained.

Haeckel says: "Even if we knew absolutely nothing of the other phenomena of development, we should be obliged to believe in the truth of the theory of descent, solely on the ground of the existence of rudimentary organs."

REPRODUCTION BY MEANS OF EGGS.

It might be thought there existed a missing link between animals which lay eggs and those which do not; this, however, is done away with in many instances—one, for example, is found in our commonest indigenous snake. The ringed snake lays eggs which require three weeks time to develop; but when it is kept in captivity, and no sand is strewn in the cage, it does not lay eggs, but retains them until the young ones are developed. This only shows how powerfully influences affect the habit of animals.

DOUBLE-SEXED INDIVIDUALS.

Another difficulty might be supposed to arise between animals which produce themselves other than by sexual reproduction. This has already been slightly touched upon; and it has been [Pg 100] shown that numerous plants and animals propagate themselves through their double-sexed organs. It occurs in a great majority of plants, but only in a minority of animals; for example, the garden-snail, leeches, earth-worms, and many other worms. Every garden-snail produces in one part of its sexual gland eggs, and in another part sperm.

Parthenogenesis offers an interesting form of transition from sexual reproduction to the non-sexual formation of germ-cells (which most resembles it). It has been demonstrated to occur in many cases among insects, especially by Seebold's excellent investigations. Among the common bees, a male individual (a drone) arises out of the eggs of the queen, if the eggs have not been fructified; a female (a queen or working bee), if the egg has been fructified.

Gonochorismus or sexual separation, which characterizes the more complicated of the two kinds of sexual reproduction, has evidently been developed from the condition of hermaphroditism at a late period of the organic history of the world. In this case the female individual in both animal and plant produces eggs or egg-cells. In animals, the male individual secretes the fructifying sperm (sperma); in plants, the corpuscles, which correspond to the sperm.

INHERITANCE.

The remarkable facts of inheritance, extending to the reproduction of unimportant peculiarities of parts or organs (rudimentary parts) mentioned above, and the occasional outbreak of ancestral characters that have been dormant through several generations (some of which I will mention further on), might be thought perfectly unexplainable; but they are readily accounted for by the supposition that each part of an organism contributes its constituent and effective molecules to the germ and sperm [Pg 101] particles. Mr. Sorby made numerous investigations with relation to the number of molecules in the germinal matter of eggs, and the spermatic matter supplied by the male. Omitting the alkali, Mr. Sorby takes the formula, $C_{72}H_{112}N_{18}SO_{22}$, as representing the composition of albumen. In a $1/2000$ of an inch cube, he reckons —

Albumen	18,000,000,000,000	molecules.
Water	992,000,000,000,000	"
	1,010,000,000,000,000	molecules.

Or, in a sphere of the same diameter, 530,000,000,000,000 of the two components. Taking a single mammalian spermatozoon, having a mean diameter of $1/6000$ of an inch; "it might contain two and a half million of such gemmules. If these were lost, destroyed, or fully developed at the rate of one in each second, this number would be exhausted in about one month; but since a number of spermatozoa appears to be necessary to produce perfect fertilization, it is quite easy to understand that the number of gemmules introduced into the ovum may be so great that the influence of the male parent may be very marked, even after having been, as regards particular character, apparently dormant for many years." The germinal vesicle of a mammalian ovum being about $1/1000$ of an inch, mean diameter, might contain five hundred million of gemmules, which, if used up at the rate of one per second, would last more than seventeen years. If the whole ovum, about $1/150$ in diameter, were all gemmules, the number would be sufficient to last, at this rate, one per second for 5,600 years! This, however, is not probable; but Mr. Sorby's remarks has completely removed all doubt as to its physical possibility from the Darwinian theory; "and they prompt us," says Slack, "to a wonderful conception of the powers residing in minute quantities of matter."

[Pg 102] The laws of inheritance are divisible into two series, conservative and progressive transmission; the laws of adaptation to direct (active) or indirect (potential) adaptation.

External causes often influence the reproductive system, especially in organism propagating in a sexual way. This can be strikingly shown in artificially produced monstrosities. Monstrosities can be produced by subjecting the parental organism to certain extraordinary conditions of life; and curiously enough, such an extraordinary condition of life does not produce a change of the organism itself, but a change in its descendants. The new formation exists in the parental organism only as a possibility (potential); in the descendants it becomes a reality (actual). Most commonly, monstrosities with very abnormal forms are sterile, but there are instances where they reproduce their kind and become a species. [35] Geoffroy St. Hilaire, who perhaps made the deepest investigations ever conducted into the nature and causes of their production, first conceived the idea of artificially producing them, and to this end he began modifications of the physical conditions of the evolution of the chicken during natural and artificial incubation. He determined the fact that monsters could be produced in this way, but scarcely carried his investigation further. This work has been taken up by M. Dareste, and he has lately published a volume in Paris which recounts the results of a quarter of a century's experimenting. Eggs, he states, were submitted to incubation in a vertical instead of a horizontal position; they were covered with varnish in certain places so as to stop or modify evaporation and respiration. The evolution of the chick was rendered slower by a temperature below that of the normal heat of incubation. Finally, eggs were warmed only at one point, so that the young animal, during development, was submitted at different parts to variable temperatures.

[Pg 103]

[Pg 104]

Fig. 1.   Fig. 2.

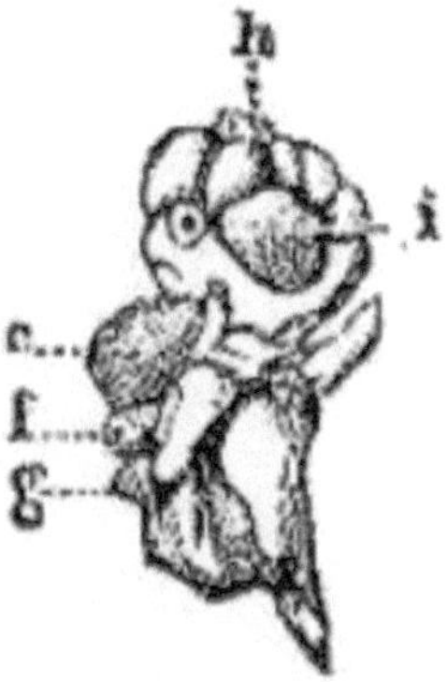

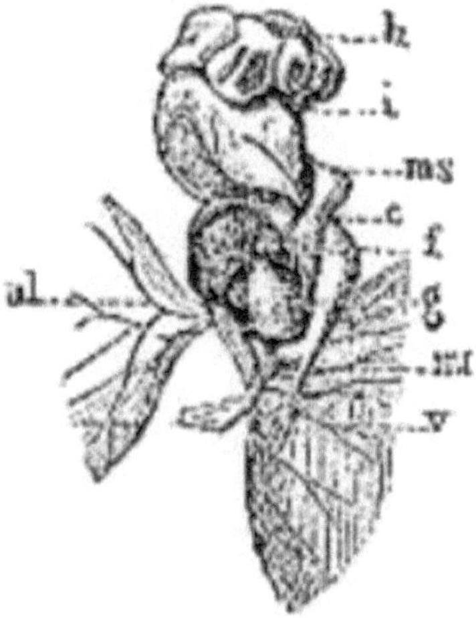

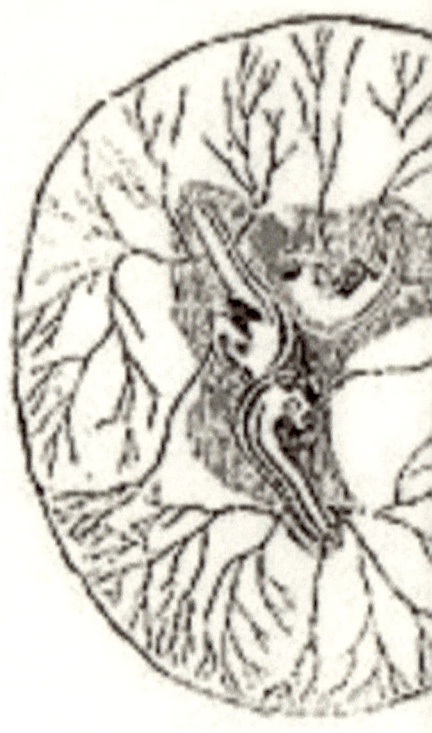

Fig. 5.
Fig. 6.　　　　　　　　　　　　Fig. 7.

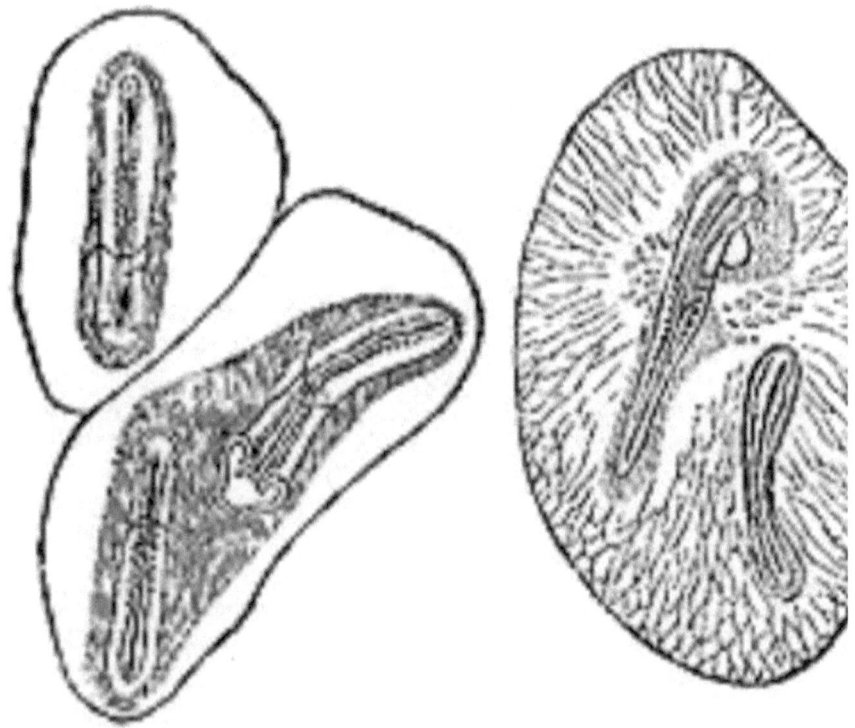

[Pg 105] These perturbations resulted in the most curious and unlooked for deformities in the embryo, some being not alone peculiar to the bird, but being similar to those which have been recognized in many other animals, and even in the human species. The data obtained have been deemed so important that M. Dareste has recently received the Lacaze prize for physiology from the French Academy of Sciences.

It would be impossible to review even a fraction of the many forms of monstrosities which M. Dareste has discovered. Those that we give will, however, suffice to convey an idea of the wonderful variations produced. Fig. 1 is a chick embryo with the encephalon entirely outside the head, the heart, liver, and gizzard outside the umbilical opening, right wing lifted up beside the head, and the development of the left one stopped. In Fig. 2 the encephalon is herniated and marked with blood spots, the eye is rudimentary and replaced by a spot of pigment, the upper beak is shorter than the lower one, while the heart, liver, etc., are all outside. In Figs. 3 and 4 the head is compressed, eyes well developed, but in the back instead of in the sides of the head; the body is bent, abdominal intestines not closed, heart largely developed and herniated. The literal references to the foregoing are: *am*, amnion; *al*, allantois; *v*, vitellus; *h*, encephalon; *i*, eye; *c*, heart; *f*, liver; *g*, gizzard; *ms*, upper, and *mi*, lower member.

The commonest case of monstrosity observed by M. Dareste has been that of the head protruding from the navel, and the heart or hearts above the head. This is a most extraordinary and new monster, and, if it persist, a chicken with its heart on its back, like a hump, may be expected. A curious fact discovered is the duplicity of the heart at the beginning of incubation, [Pg 106] two hearts, beating separately, being clearly seen. Another anomaly consists in heads with a frontal swelling, which is filled by the cerebral hemispheres.

M. Dareste's artificial monsters are all produced from the single germ or cicatricule (as the white circular spot seen in the yellow of the egg, and from which the embryo springs, is termed). He has not yet been able to determine artificially the production of monsters, the origin of which takes place in a peculiar state of the cicatricule before incubation. But having submitted to incubation some 10,000 eggs, he has obtained several remarkable examples of double monstrosities in process of formation, some representations of which are given herewith. Fig. 5 shows three embryos, all derived from a single cicatricule. Fig. 6 represents three embryos from two cicatricules. On one side of the line of junction are two imperfectly developed embryos, one having no heart. The single embryo on the other side is generally normal, but has a heart on the right side. In Fig. 7 are twins, one well formed, the heart circulating colorless blood, the other having no heart and a rudimentary head. Fig. 8 exhibits a double monster with lateral union. The heads are separate, and there are three upper and three lower members, those of the latter on the median line belonging equally to each of the pair.

ACQUIRED QUALITIES.

When an organism has been subjected to abnormal conditions in life it can transmit any peculiarity it may have acquired. This is, however, not always possible, otherwise descendants of men who have lost their arm or leg would be born without the corresponding arm or leg—this shows that some acquired qualities are more easily transmitted than others—although there are cases, as, for instance, a race of dogs without tails has [Pg 107] been produced by cutting off the tails of both sexes of the dog, during several generations. "A few years ago," says Haeckel, "a case occurred on an estate near Jena in which, by the careless slamming of a stable-door, the tail of a bull was wrenched off, and the calves begotten by this bull were all born without a tail. This is certainly an exception; but it is very important to note the fact that under certain unknown conditions such violent changes are transmitted in the same manner as many diseases." The transmission of diseases such as consumption, madness, and albinism form examples. Albinoes are those individuals who are distinguished by the absence of coloring matter from their skins; they are of frequent occurrence among men, animals and plants. Among many animals, such as rabbits and mice, albinoes with white fur and red eyes are so much liked that they are propagated. This would be impossible were it not for the law of the transmission of adaptations. Hornless cattle have descended from a single bull born in 1770 of horned parents, but whose absence of horns was the result of some unknown cause.

The law of interrupted or latent transmission, as illustrated in grandchildren who are like the grandparents, but quite unlike the parents. Animals often resume a form which have not existed for many generations. One of the most remarkable instances of this kind of reversion, or "atavism," is the fact that in some horses there sometimes appear singular dark stripes similar to those of the zebra, quagga, and other wild species of African horse.

Nutrition directly modifies adaptation, as is well illustrated by animals which have been bred for domestic or other purposes. If a farmer is breeding for fine wool he gives much different food to the sheep than he would if he wished to obtain flesh or an abundance of fat. Even the bodily form of man is quite different according to its nutrition. Food containing [Pg 108] much nitrogen produces little fat, that containing little nitrogen produces a great deal of fat. People who by means of Banting's system, at present so popular, wish to become thin, eat only meat and eggs—no bread, no potatoes.

Man can breed for milk in cattle, for feathers in pigeons, for colored flowers in plants, and, in fact, for almost any desirable quality.

GEOLOGICAL RECORD.

The Geological Record (palæontology) furnishes weighty evidence of man's descent; for the circumstantial evidence derived from this source is written without the possibility of a mistake, with no chance of error, on the stratified rocks. It is true that the geological record must be incomplete, because it can only preserve remains found in certain favorable localities, and under particular conditions; that this valuable record must be destroyed by processes of denudation, and obliterated by processes of metamorphosis, it cannot be doubted. "Beds of rock of any thickness, crammed full of organic remains, may yet," says Huxley, "by the percolation of water through them, or the influence of subterranean heat (if they descend far enough toward the centre of the earth), lose all trace of these remains, and present the appearance of beds of rock formed under conditions in which there was no trace of living forms. Such metamorphic rocks occur in formations of all ages; and we know with perfect certainty, when they do appear, that they have contained organic remains, and that those remains have been absolutely obliterated." If we look at the geological record, we find:

The First Epoch. — *The Archilithic*, or Primordial Epoch, constitutes the *Age of Skull-less Animals and Sea-weed Forests*, and is made up of the Laurentian, Cambrian, and Silurian Period.

The Second Epoch. — *The Palæolithic*, or Primary Epoch, [Pg 109] constitutes the *Age of Fishes and Fern Forests*, and is made up of the Devonian, Coal, and Permian Period.

The Third Epoch. — *The Mesolithic*, or Secondary Epoch, constitutes the *Age of Reptiles and Pine Forests, Coniferæ*, and is made up of the Triassic, Jurassic, and Chalk Period.

The Fourth Epoch. — *The Cænolithic*, or Tertiary Epoch, constitutes the *Age of Mammals and Leaf Forests*, and is made up of the Eocene, Miocene, and Phocene Period.

The Fifth Epoch. — The *Anthropolithic*, or Quaternary Epoch, constitutes the *Age of Man and Cultivated Forests*, and is made up of the Glacial and Postglacial Period, and the Period of Culture.

During the archilithic epoch the inhabitants of our planet, as has been already stated, consisted of skull-less animals, or aquatic forms. No remains of terrestrial animals or plants, dated from this period, have as yet been found.

The archilithic period was longer than the whole long period between the close of the archilithic and the present time; for if the total thickness of all sedimentary strata be estimated as about one hundred and thirty thousand feet, then seventy thousand feet belong to this epoch. It was during this epoch that the little mass of protoplasm, which has been so often spoken of, came into existence.

It has been stated above that palæontology is quite deficient. This is not only true of the record, but of the lack as yet of sufficient investigations. The greatest fields of investigation in this department have never been explored. The whole of the petrifactions accurately known do not probably amount to a hundredth part of those which, by more elaborate explorations, are yet to be discovered. The most ancient of all distinctly preserved petrifactions is the Eozoon Canadense, which was found in the lowest Laurentian strata in the Ottawa formation.

[Pg 110] Probably no discovery in palæontology ranks higher than the discovery of the descendants of the horse. The horse, for example, as far as his limbs and teeth go, differs far more from extant graminivora than man differs from the ape. Had not fossil ungulates been found, which demonstrate the common origin of the horse with didactyles and multidactyles, some would have deemed the horse a special miraculous creation. But now the links are complete, and the descent of the horse is found to follow exactly what the doctrine of evolution could have predicted.

ONTOGENY.

It has been stated that the palæontological record is quite incomplete, owing to many facts, some of which have been mentioned; fortunately, the history of the development of the organic individual, or ontogeny, comes in to fill up many deficiencies.

Ontogeny is a repetition of the principal forms through which the respective individuals have passed from the beginning of their tribe, and its great advantage is that it reveals a field of information which it was impossible for the rocks to retain; for the petrification of the ancient ancestors of all the different animal and vegetable species, which were soft, tender bodies, was not possible.

The annexed plate illustrates the dog, rabbit, and man in their first stages of development. Illustrations of a fish, an amphibious animal, a reptile, a bird, or any mammal, could also be given; for all vertebrate animals of the most different classes, in their early stages of development, cannot be distinguished, and the nearer the animal approaches man in the ascending scale, the longer does this similarity continue to exist—when reptiles and birds are distinctly different from mammals, the dog and the man are almost identical.

The gill-arches of the fish exist in man, in dogs, in fowls, in reptiles, and in other vertebrate animals during the first stages of their development. Man also possesses, in his first stages, a real tail, as well as his nearest kindred—the tailless apes (orang-outang, chimpanzee, gorilla), and vertebrate animals in general. The tail, as has been stated, man still retains, though hidden as a rudiment.

[Pg 111]

Fig. I.   Fig. I'

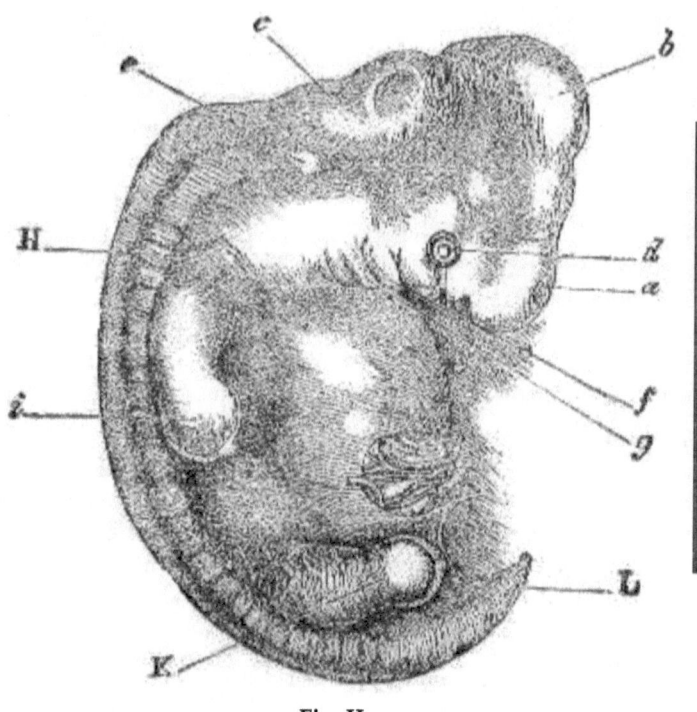

Fig. II.

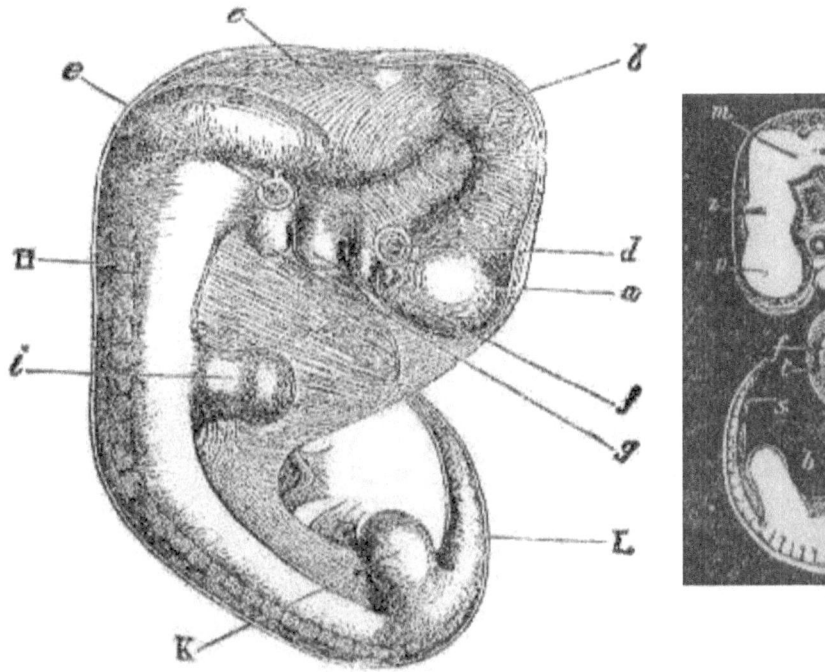

Fig. III.   Fig. V

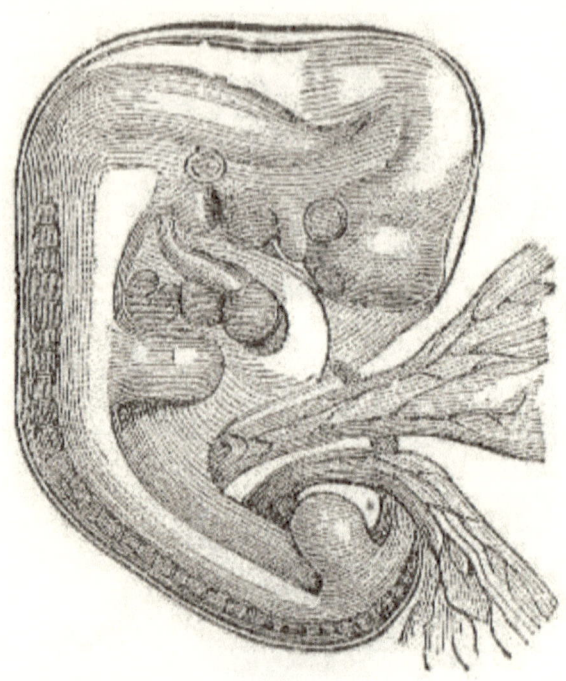

Fig. I.—Human Embryo.—*Ecker*.
Fig. II.—Embryo of Dog.—*Bischoff*.
Fig. III.—Dog Embryo.—*Huxley*.
Figs. IV, V, and VI.—Embryo of Rabbit in three stages of development.—*Haeckel*.
Figs. VII, VIII, and IX.—Embryo of Man in three stages of development.—*Haeckel*. v, fore brain; z, twix brain; m, middle brain; h, hind brain; n, after brain; r, spinal marrow; e, nose; a, eye; o, ear; k, gillarches; g, heart; w, vertebral column; f, fore limbs; b, hind limbs; s, tail.

[Pg 112]

[Pg 113] "Man presents in his earliest stages of embryonic growth, a skeleton of cartilage, like that of the lamprey; also, five origins of the aorta and five slits on the neck, like the *lamprey* and the *shark*. Later, he has but four aortic origins, and a heart now divided into two chambers, like *bony fishes*; the optic lobes of his brain also having a very fish-like predominance in size. Three chambers of the heart and three aortic origins follow, presenting a condition permanent in the *batrachia*; then two origins with enlarged hemispheres of the brain, as in *reptiles*. Four heart chambers and one aortic root on each side, with slight development of the cerebellum, agree with the characters of the *crocodiles*, and immediately present the special mammalian conditions, single aortic root, and the full development of the cerebellum. Later comes that of the cerebrum, also in its higher mammalian or human traits." At no time in the development of the egg, save at the start, do the embryos of the various vertebra assume the *exact* or *entire* characteristics of one another, but they assimilate so closely that it requires the eye of the expert to distinguish them; and, as has already been stated, the more closely an animal resembles another, the longer and the more intimately do their embryos resemble one another; so that, for example, the embryo of the snake and of a lizard remain like one another longer than do those of a snake and of a bird; and the embryo of a dog and of a cat remain like one another for a far longer period than do those of a dog and a bird, or a dog and an opossum, or even those of a dog and a monkey.

[Pg 114] Surely it must be admitted that the short brief history given by the development of the egg, is far more wonderful than phylogeny or the long and slow history of the development of the tribe, which has taken thousands of years. Compare this time with the time required for the development of the smallest mammals— the harvest mice which develops in three weeks, or the smallest of all birds, the humming-bird, which quits the egg on the twelfth day, or with man who passes through the whole course of his development in forty weeks, or with the rhinoceros who requires 1½ years, or the elephant who requires ninety weeks. How insignificant are these various periods to the long period originally required; yet in these short periods the whole phylogeny is run through in the ontogeny or the history of the development of the egg.

[Pg 115]

THE ATTRIBUTES OF MAN.

We must now consider briefly some of the attributes of man, and see if he really possesses attributes which are in no inferior degree possessed by animals. Before proceeding directly to the consideration of the attributes of man, it will be best to show the correlation that exists between what are called man's vital forces and the physical forces of nature. To do this let us choose three forms of its manifestation: these shall be heat evolved within the body; muscular energy or motion; and lastly, nervous energy or that form of force which, on the one hand, stimulates a muscle to contract, and on the other appears in forms called mental. It will not take any extensive argument to demonstrate that the heat of the body does not differ from heat from any other source. It is known that the food taken into the body contains potential energy, which is capable of being in part converted into actual heat by oxidation; and since we know that the food taken into the body is oxidized by the oxygen of the air supplied by the lungs, the heat of the body must be due to the slow oxidation of the carbon, perhaps also hydrogen, sulphur, and phosphorus in the food. Now since this so-called vital heat is developed by oxidation, is recognized by the same tests and applied to the same purposes as any other heat, it is as truly correlated to the other forces as when it has a purely physical origin. The amœboid activity of a white blood corpuscle is stimulated within certain limits by heat. [Pg 116] Hatching of eggs and the germination of seeds may be likewise hastened or retarded by access or deprivation of heat. It was considerations such as these which led to the doctrine of correlation of the vital and physical forces.

With respect to the muscular force exerted by an animal, it was supposed that it was created by the animal. Dr. Frankland [36] says to this: "An animal can no more generate an amount of force capable of moving a grain of sand, than a stone can fall upwards or a locomotive drive a train without fuel." As the amount of CO_2 exhaled by the lungs is increased in the exact ratio of work done by the muscle, it cannot be doubted that the actual force of the muscle is due to the converted potential energy of the food. Since every exertion of a muscle and nerves involves the death and decay of those tissues to a certain extent, as shown by the excretions, Prof. Orton [37] has been led to say: "An animal begins to die the moment it begins to live." "A muscle," says Barker, [38] "is like a steam-engine, is a machine for converting the potential energy of carbon into motion; but unlike a steam-engine, the muscle accomplishes this conversion directly, the energy not passing through the intermediate stages of heat. For this reason the muscle is the most economical producer of mechanical force known." The muscles which give the downward stroke of the wing of a bird are fastened to the breastbone, and their power in proportion to the weight of the bird is as 10,000 to 1. This great power is needed, for the air is 770 times lighter than water; the hawk being able to travel 150 miles an hour.

The last of the so-called vital forces under consideration, is that [Pg 117] produced by the nerves and nervous centres. Barker says: "In the nerve which stimulates a muscle to contract, this force is undeniably motion, since it is propagated along this nerve from one extremity to the other." This force has been likened unto electricity, the gray or cellular matter being the battery, the white or fibrous matter the conductors. Du Bois Reymond [39] has demonstrated that this force is not electricity, though by showing that its velocity is only ninety-seven feet a second. The velocity varies, though, in different animals; it is, according to Prof. Orton, [40] "more rapid in warm-blooded than in cold-blooded animals, being nearly twice as fast in man as in the frog." Wheatstone, by his method, gives the velocity of electricity in copper wire at 62,000 geographical miles per second; but as neither Fizeau, Gould, Gonnelle and others could arrive at the same result, the method was shown to be incorrect, and it remained for Dr. Siemen [41] to discover the true method, which gives the velocity just one-half that of Wheatstone's estimate, or 31,000 geographical miles per second. In the opinion of Bence Jones, the propagation of a nervous impulse is a sort "of successive molecular polarization, like magnetism." But that this agent is a force as analogous to electricity as is magnetism, is shown not only by the fact that the transmission of electricity along a nerve will cause the contraction of a muscle to which it leads, but also by the important fact discovered by Marshall, that the contraction of a muscle is excited by diminishing its normal electrical current, [42] a result which could take place only with a stimulus, says Barker, "closely allied to electricity. Nerve force must therefore be transmuted potential energy." Prof. Huxley says, [43] "the results [Pg 118] of recent inquiries into the structure of the nervous system of animals, converge toward the conclusion that the nerve-fibres which we have hitherto regarded as ultimate elements of nervous tissue, are not such, but are simply the visible aggregations of vastly more attenuated filaments, the diameter of which dwindles down to the limits of our present microscopic vision, greatly as these have been extended by modern improvements of the microscope; and that a nerve is, in its essence, nothing but a linear tract of specially modified protoplasm between two points of an organism, one of which is able to affect the other by means of the communication so established. Hence it is conceivable that even the simplest living being may possess a nervous system."

Herbert Spencer [44] says all direct and indirect evidence "justifies us in concluding that the nervous system consists of *one* kind of matter. In the gray tissue this matter exists in masses containing *corpuscles*, which are soft and have granules dispersed through them, and which, besides being thus unstably composed, are placed so as to be liable to disturbances to the greatest degree. In the white tissue this matter is collected together in extremely slender *threads* that are denser, that are uniform in texture, and that are shielded in an unusual manner from disturbing forces, except at their two extremities."

The last consideration is that form of force (thought power) which appears in manifestations called mental. It must be noticed at the outset, that every external manifestation of thought force is a muscular one, as a word spoken or written, a gesture, or an expression of the face always takes place; hence this force must be intimately correlated to nerve force. It is very certain, then, that thought force is capable in external manifestations of converting itself into actual motion. But here the question arises, can it be manifested inwardly without such [Pg 119] a transformation of energy? Or is the evolution of thought entirely independent of the matter of the brain?

This question can be answered by actual experiment, strange as it may appear. Experiments have demonstrated that any change of temperature within the skull was soonest manifested externally in that depression which exists just above the occipital protuberance.

Here Lombard [45] fastened to the head at this point two little bars, one made of bismuth, the other of an alloy of antimony and zinc, which were connected with a delicate galvanometer; [46] to neutralize the result of a gradual rise of temperature over the whole body, a second pair of bars, reversed in direction, was attached to the leg or arm, so that if a like increase of heat came to both, the electricity developed by one would be neutralized by the other, and no effect would be produced by the needle unless only one was affected. By long practice it was ascertained that a mental torpor could be induced, lasting for hours, in which the needle remained stationary. But let a person knock on the door outside of the room, or speak a single word, even though the experimenter remained absolutely passive, the reception of the intelligence caused the needle to swing twenty degrees. "In explanation of this production of heat," says Barker, [47] "the analogy of the muscle at once suggests itself. No conversion of energy is complete, and as the heat of muscular action represents force which has escaped conversion into motion, so the heat evolved during the reception of an idea is energy which has escaped conversion into thought, from precisely the same cause." Dr. Lombard's experiments have shown that the amount of heat developed by [Pg 120] the recitation to one's self of emotional poetry, was in every case less when recitation was oral; this is of course accounted for by the muscular expression. Chemistry teaches that thought-force, like muscle-force, comes from the food, and demonstrates that the force evolved by the brain, like that produced by the muscle, comes not from the disintegration of its own tissue, but is the converted energy of burning carbon. [48] "Can we longer doubt," says Barker, [49] "that the brain too, is a machine for the conversion of energy? Can we longer refuse to believe that even thought force is in some mysterious way correlated to the other natural forces? and this even in the face of the fact that it has never yet been measured. [50] Have we not a right to ask 'why a special force (vital force) should be needed to effect the transformation of physical forces into those modes of energy which are active in the manifestation of living beings, while no peculiar force is deemed necessary to effect the transformation of one mode of physical force into any other mode of physical force?"

Richard Owen says: [51] "In the endeavor to clearly comprehend and explain the functions of the combination of forces called 'brain,' the physiologist is hindered and troubled by the views of the nature of those cerebral forces which the needs of dogmatic theology have imposed on mankind. *** Religion, pure and undefiled, can best answer how far it is righteous or just to charge a neighbor with being unsound in his principles who holds the term 'life' to be a sound expressing the sum of [Pg 121] living phenomena, and who maintains these phenomena to be modes of force into which other forms of force have passed from potential to active states, and reciprocally, through the agency of the sums or combinations of forces impressing the mind with the ideas signified by the terms 'monad,' 'moss,' 'plant,' or 'animal.'"

We have now shown that the very forces which give vent to the attributes of man, are correlated to the physical forces. Let us now consider his attributes as manifested by his mental powers. There is no doubt the difference between the mental faculties of the ape and that of the lowest savage, who cannot express any number higher than four and who uses hardly any abstract terms for common objects or for the affections, [52] is still very great and would still be great, says Darwin, "even if one of the higher apes had been improved or civilized as much as a dog has been in comparison with its parent form, the wolf or jackal." But when we examine the interval of mental power between one of the lowest fishes, as a lamprey or a lancelet, and one of the higher apes, and recognize the fact that this interval is filled up by numberless gradations, it does not become so difficult to understand the interval between an ape and man, which is not by far so great. As in finding out what is peculiar to a living body in distinction to a body not living, we found it absurd to take man as the perfection of the animal scale—the microscopic monad possessing life as well as him—so in the case of man's mental attributes, which have always been increasing, always perfecting, since the first genuine man came into existence, it would be equally absurd to compare the intellectual man of to-day with an ape to see what attributes he possesses which the ape does not possess; but if we go down in the scale and compare the savage with the ape, the difficulty is not by far so [Pg 122] great. It will be found on close examination, though, that man and the higher animals, especially the primates, have many instincts in common. "All," says Darwin, "have the same senses, intuitions and sensations; similar passions, affections, and emotions; even the more complex ones, such as jealousy, suspicion, emulation, gratitude and magnanimity; they practice deceit and are revengeful; they are sometimes susceptible to ridicule and even have a sense of humor; they feel wonder and curiosity; they possess the same faculties of imitation, attention, deliberation, choice, memory, imagination, the association of ideas, and reason, though in very different degrees. The individuals of the same species graduate in intellect from absolute imbecility to high excellence; they are also liable to insanity, though far less often than in the case of man." [53] Nevertheless, in the face of these facts, many authors have insisted that man is divided by an inseparable barrier from all the lower animals in his mental faculties. It only shows the improper or imperfect consideration of the subject they have under discussion.

It may be thought at first that some of the mental attributes mentioned above are not possessed by animals. I therefore will briefly consider a few of the more complex ones. We can dismiss the consideration of such attributes as happiness, terror, suspicion, courage, timidity, jealousy, shame, and wonder, as well-known attributes. *Curiosity* in animals is often observed. An instance mentioned by Brehm will serve to illustrate: Brehm gives a curious account of the instinctive dread which his monkeys exhibited for snakes; but their curiosity was so great that they could not desist from occasionally satiating their horror in a most human fashion, by lifting up the lid of the box in which the snakes were kept. *Imitation* is also found among the action of [Pg 123] animals, especially among monkeys, which are well known to be ridiculous mockers.

It is unnecessary to refer to the faculty of attention, as it is common to almost all animals, and the same may be said of memory as for persons or places.

One would hesitate to believe an animal possesses *imagination*, but such is the case. Dreaming, it will be admitted, gives us the best notion of this power. Now as dogs, cats, horses, and probably all the higher animals, even birds, have vivid dreams—this is shown by their movements and the sounds uttered—"we must admit," says

Darwin, "they possess some power of imagination. There must be something special which causes dogs to howl in the night, and especially during moonlight, in that remarkable and melancholy manner, called baying. All dogs do not do so; and, according to Housyeau, [54] they do not look at the moon, but at some fixed point near the horizon. Housyeau thinks that their imaginations are disturbed by the vague outlines of the surrounding objects, and conjure up before them fantastic images; if this be so, their feelings may almost be called superstitious."

The next mental faculty is *reason*, which stands at the summit; but still there are few persons who will deny that animals possess some power of reasoning. A few illustrations will be all that is necessary to satisfy the inquiring mind on this point. Reugger, a most careful observer, states that when he first gave eggs to his monkey in Paraguay they smashed them, and thus lost much of their contents; afterward they gently hit one end against some hard body, and picked off the bits of shell with their fingers. After cutting themselves *once* with any sharp tool, they would not touch it again, or would handle it with the greatest caution. Lumps of sugar were often given them, wrapped [Pg 124] up in paper; and Reugger sometimes put a live wasp in the paper, so that in hastily unfolding it they got stung; after this had *once* happened, they afterward first held the packet to their ears to detect any movement within.

The following cases relating to dogs are described by Darwin: Mr. Colquhoun winged two wild ducks, which fell on the farther side of a stream; his retriever tried to bring over both at once, but could not succeed; she then, though never before known to ruffle a feather, deliberately killed one, brought over the other, and returned for the dead bird. Colonel Hutchinson relates that two partridges were shot at once—one being killed, the other wounded; the latter ran away, and was caught by the retriever, who, on her return, came across the dead bird; "she stopped, evidently greatly puzzled, and after one or two trials, finding she could not take it up without permitting the escape of the winged bird, she considered a moment, then deliberately murdered it by giving it a severe crunch, and afterward brought away both together. This was the only known instance of her ever having wilfully injured any game. Here we have reason, though not quite perfect; for the retriever might have brought the wounded bird first, and then returned for the dead one, as in the case of the two wild ducks. I give the above cases as resting on the evidence of two independent witnesses; and because in both instances the retrievers, after deliberation, broke through a habit which was inherited by them (that of not killing the game retrieved), and because they show how strong their reasoning faculty must have been to overcome a fixed habit." [55]

It has often been said that no animal uses any tool, but this can be so easily refuted on reflection, that it is hardly worth while considering; for illustration, though, the chimpanzee in a state of nature cracks nuts with a stone; Darwin saw a young orang put a [Pg 125] stick in a crevice, slip his hand to the other end, and use it in a proper manner as a lever. The baboons in Abyssinia descend in troops from the mountains to plunder fields, and when they meet troops of another species a fight ensues. They commence by rolling great stones at their enemies, as they often do when attacked with fire-arms.

The Duke of Argyll remarks that the fashioning of an implement for a special purpose is absolutely peculiar to man; and he considers this forms an immeasurable gulf between him and the brutes. "This is no doubt," says Darwin, "a very important distinction; but there appears to me much truth in Sir J. Lubbock's suggestion, [56] that when primeval man first used flint-stones for any purpose, he would have accidentally splintered them, and would then have used the sharp fragments. From this step it would be a small one to break the flints on purpose, and not a very wide step to fashion them rudely. The later advance, however, may have taken long ages, if we may judge by the immense interval of time which elapsed before the men of the neolithic period took to grinding and polishing their stone tools. In breaking the flints, as Sir J. Lubbock likewise remarks, sparks would have been emitted, and in grinding them heat would have been evolved; thus the two usual methods of 'obtaining fire may have originated.' The nature of fire would have been known in many volcanic regions where lava occasionally flows through forests."

It becomes a difficult task to determine how far animals exhibit any traces of such high faculties as *abstraction, general conception, self-*

consciousness, mental individuality. There can be no doubt, if the mental faculties of an animal can be improved, that the higher complex faculties such as abstraction and self-consciousness have developed from a combination of the simpler ones; this seems to be well illustrated in the young child, as [Pg 126] such faculties are developed by imperceptible degrees. These high faculties are very sparingly possessed by the savage; as Buchner [57] has remarked, how little can the hard-worked wife of a degraded Australian savage, who uses very few abstract words and cannot count above four, exert her self-consciousness or reflect on the nature of her own existence. If there exist a class of people so inferior in their mental faculties as these, it is not difficult for us to understand how the educated animal who possesses memory, attention, association, and even some imagination and reason, can become capable of abstraction, &c., in an inferior degree even to the savage. It certainly cannot be doubted that an animal possesses mental individuality—as when a master returns to a dog which he has not seen for years, and the dog recognizes him at once.

One of the chief distinctions between man and animals is the faculty of language. Let us look at this for a moment. "The essential differences," says Prof. Whitney, "which separate man's means of communication in kind as well as degree from that of the other animals is that, while the latter is instinctive, the former is in all its parts arbitrary and conventional. No man can become possessed of any language without learning it; no animal (that we know of) has any expression which he learns, which is not the direct gift of nature to him." Any child of parents living in a foreign country grows up to speak the foreign speech, unless carefully guarded from doing so; or it speaks both this and the tongue of its parent with equal readiness. A child must learn to observe and distinguish before speech is possible, and every child begins to know things by their name before he begins to call them. "If it were not for the added push," says Prof. Whitney, "given by the desire of communication, the great and wonderful [Pg 127] power of the human soul would never move in this particular direction; but when this leads the way, all the rest follows." No philologist now supposes that any language has been deliberately invented; it has been slowly and unconsciously developed by many steps.

There can be no question that language owes its origin to the imitation and modification of various natural sounds, the voices of other animals, and man's own instinctive cries, aided by signs and gestures; and this is the opinion of Max Müller. And Prof. Whitney remarks that "spoken language began, we may say, when a cry of pain, formally wrung out by real suffering, and seen to be understood and sympathized with, was repeated in imitation, no longer as a mere instinctive utterance, but for the purpose of intimating to another." Darwin says that "the early progenitor of man probably first used his voice in producing true musical cadences, that is, in singing, as do some gibbon-apes at the present day. It is therefore probable that the imitation of musical cries by articulate sounds may have given rise to words expressive of very complex emotions."

The nearest approach to language are the sounds uttered by birds. All that sing exert their power instinctively, but the actual song, and even the call notes, are learned from their parents or foster-parents. These sounds are no more innate than language is in man, as has been proved by Davies Barrington. [58] The first attempt to sing "may be compared to the imperfect endeavor in a child to babble." Prof. Whitney says, if the last transition forms of man "could be restored, we should find the transition forms toward our speech to be, not at all a minor provision of natural articulate signs, but an inferior system of conventional signs, in tone, gesture, and grimace. As between these three natural means of expression, it is simply by a kind of process of [Pg 128] natural selection and survival of the fittest that the voice has gained the upper hand, and come to be so much the most prominent that we give the name of language (tonguiness) to all expression." A single utterance or two at first had to do the duty of a whole clause; afterward man learned to piece together parts of speech, and thus arose sentences.

Although no language, as has already been said, has been deliberately invented, "still each word may not be unfitly compared to an invention; it has its own place, mode, and circumstances of devisal, its preparation in the previous habits of speech, its influence in determining the after progress of speech development; but every language in the gross is an institution, on which scores or hundreds of generations and unnumbered thousands of individual workers have labored." [59]

There is no question at all but that the mental powers in the earliest progenitors of man must have been more highly developed than in the ape, before even the most imperfect form of speech could have come into use; but the constant advancement of this power would have reacted on the mind to enable it to carry on longer trains of thought. "A complex train of thought," says Darwin, "can no more be carried on without the aid of words, whether spoken or silent, than a long calculation without the use of figures in algebra. It appears also that even an ordinary train of thought almost requires or is greatly facilitated by some form of language; for the dumb, deaf, and blind girl, Laura Bridgman, was observed to use her fingers while dreaming. [60] Nevertheless a long succession of vivid ideas may pass through the mind, without the aid of any form of language, as we may infer from the movements of dogs during their dreams."

The struggle for existence is going on in every language; one after another will be swept out of existence, and the languages best fitted for the practical uses of the masses of people will alone survive. Max Müller has well remarked: "A struggle for life is constantly going on amongst the words and grammatical forms in each language. The better the shorter; the easier forms are constantly gaining the upper hand, and they owe their success to their own inherent virtue." [61]

It must not be thought for a moment that that which distinguishes a man from the lower animals is the understanding of articulate sounds—for, as every one knows, dogs understand many words and sentences; and Darwin says, at this stage they are at the same stage of development as infants, between the ages of ten and twelve months, who understand many words and sentences, but still cannot utter a single word. It is not the mere articulation which is our distinguishing character; for parrots and other birds possess the power. Nor is it the mere capacity of connecting definite sounds with definite ideas; for it is certain that some parrots, which have been taught to speak, connect unerringly words with things, and persons with events." The lower animals, as has already been stated, differ from man solely in his almost infinitely larger power of associating together the most diversified sounds and ideas; and this obviously depends on the high development of his mental powers.

We now come to the consideration of a very delicate subject—a subject which is certainly at best very unsatisfactory to handle, as far as popular sentiment is concerned; for, no matter how successfully it may be handled, according to one class of thinkers, to another class of more orthodox thinkers it would be entirely at fault. The subject is, *Man's Moral Sense, Belief in God, Religion, Conscience, and Hope of Immortality.*

It has been stated by some writers that where "faith com [Pg 130] mences science ends." How erroneous is such a statement as this! for, as Krauth has said, "The great body of scientific facts is actually the object of knowledge to a few, and is supposed to be a part of the knowledge of the many, only because the many have faith in the statements of the few, though they can neither verify them, nor even understand the processes by which they are reached." [62]

"We believe," says Lewes, "that the sensation of violet is produced by the striking of the ethereal waves against the retina more than seven hundred billions of times in a second. * * * These statements are accepted *on trust* by us who know that there are thinkers for whom they are irresistible conclusions." It is evident that it is to faith that science owes, to a very great extent, her progress and development; for it is impossible for man to prove by experimental demonstration all the facts of science, and since a certain number of facts have got to be accepted before a new experiment can be attempted, he has to accept on faith that such and such a statement is a fact, because such and such a scientist has claimed to have demonstrated it. "We are not *responsible* for the fact," says Krauth, "that under the conditions of knowledge we *know*, or in defect of them do not know; we are responsible if, under the conditions of a well-grounded faith, we disbelieve." [63]

Let us look, then, at the belief in God. The question under consideration at first will not be whether there exists a God, the creator and ruler of the universe—for this will be afterward considered—but is there any evidence that man was aboriginally endowed with the ennobling belief in the existence of an Omnipotent God.

Schweinfurth relates that the Niam-niam, that highly inter [Pg 131] esting dwarf people of Central Africa, have no word for God, and therefore, it must be supposed, no idea; and Moritz Wagner has given a whole selection of reports on the absence of religious consciousness in inferior nations. The idea that conscience is a sort of permanent inspiration or dwelling of God in the soul, I think, on consideration, any reasonable man will not assume. "It is a purely human faculty," says Savage, "like the faculty for art or music; and it gets its authority, as they do by being true, and just in so far as it is true. Consciousness is our own knowledge of ourselves and of the relation between our own faculties and powers. Conscience is our recognition of the relations, as right or wrong, in which we stand to those about us, God and our fellows. *Con-scio* is to know with, in relation.

There is such a thing, of course, as a *false conscience* and a *true conscience*. All the false "conscientiousness grows out of the fact that men suppose they stand in certain relationships that do not really exist. Thus they imagined duties that are not duties at all." The virtues which must be practised by rude men, so that they can hold together in tribes, are of course important. No tribe could hold together if robbery, murder, treachery, etc., were common; in other words, there must be honor among thieves. "A North-American Indian is well pleased with himself, and is honored by others, when he scalps a man of another tribe; and a Dyak cuts off the head of an unoffending person, and dries it as a trophy. The murder of infants has prevailed on the largest scale throughout the world, and has been met with no reproach; but infanticide, especially of females, has been thought to be good for the tribe, or at least not injurious. Suicide during former times was not generally considered as a crime, but rather, from the courage displayed, as an honorable act; and it is still practised by some semi-civilized and savage nations without reproach, for it does not obviously concern others [Pg 132] of the tribe. It has been recorded that an Indian Thug conscientiously regretted that he had not robbed and strangled as many travelers as did his father before him." [64]

See how weak the conscience of even more highly civilized men are in their dealings with the brute creation; how the sportsman delights in hunting-scenes, Spanish bull-fights, cock-fights, etc.;

how indignant was the sensitive Cowper, if any one should "needlessly set foot upon a worm"! The rights of the worm are as sacred in his degree as ours are, and a true conscience will recognize them. What, then, is a true conscience? Savage states in a few words, it is "one that knows and is adjusted to the realities of life. When men know the truth about God, about themselves—body and mind and spirit—about the real relations of equity in which they stand to their fellow-men in state and church and society, and when they appreciate these, and adjust their conscience to them, then they will have a true conscience. An absolutely true conscience, of course, cannot exist so long as our knowledge of the reality of things is only partial."

It is evident, then, that the conscience of man depends on his education and environments, and therefore is the subject of improvement. It becomes, then, the duty of every man to search for truth, for his conscience is not infallible, and by so doing he will bring it to accord with the real facts of God. "Throw away," says Savage, "prejudice and conceit, seek to make your conscience like the magnetic needle. The needle ever and naturally seeking the unchanging pole." As conscience, then, is but a faculty capable of development, it is not so difficult to understand a race of people whose conscience was in just the first stages of development; and, finally, a race which did not possess this faculty at all, as in the inferior nations which Wagner speaks of.

[Pg 133]

[Pg 134]

Fig. I. — Butcher's Shop of the Anziques, Anno 1598.
(From Man's Place in Nature, by *Huxley*.)

[Pg 135] What kind of conscience and intelligence had the people near Cape Lopez, called the Anziques, which M. du Chaillu describes. They had incredible ferocity; for they ate one another, sparing neither friends nor relations. Their butcher-shops were filled with human flesh, instead of that of oxen or sheep, for they ate the enemies they captured in battle. They fattened, slayed, and devoured their slaves also, unless they thought they could get a good price for them; and moreover, for weariness of life or desire for glory (for they thought it a great thing and a sign of a generous soul to despise life), or for love of their rulers, offered themselves up for food. There were, indeed, many cannibals, as in the East Indies and Brazil and elsewhere, but none such as these, since the others only ate their enemies, but these their own blood relations.

There is therefore, combining the fact mentioned by Wagner with the fact that some nations have no idea of one or more gods, not even a word to express it (proving that they have no idea), I say, there is therefore no evidence that man was aboriginally endowed with any such belief as the existence of an Omnipotent God; and in this assertion almost all the learned men concur. "If, however," says Darwin, "we include under the term religion, the belief in unseen or spiritual agencies, the case is wholly different; for this belief seems to be universal with the less civilized races. Nor is it difficult to understand how it arose."

The savage has a stronger belief in bad spirits than in good ones. "The same high mental faculties which first led man to believe in unseen spiritual agencies, then in fetishism, polytheism, and ultimately in monotheism, would infallibly lead him, as long as his reasoning powers remained poorly developed, to very strange superstitions and customs. Many of these are terrible to think of: such as the sacrifice of human beings to a blood- [Pg 136] loving god, the trial of innocent persons by the ordeal of poison, of fire, of witchcraft, etc.; yet it is well occasionally to reflect on these superstitions, for they show us what an infinite debt of gratitude we owe to the improvement of our reason, to science, and to our accumulated knowledge." [65] As Sir J. Lubbock has well observed: "It is not too much to say that the possible dread of unknown evil hangs like a thick cloud over savage life, and embitters every pleasure. These miserable and indirect consequences of our highest faculties may be compared with the incidental and occasional mistakes of the instincts of the lower animals."

The belief, then, of the existence of an Omnipotent God came with the development of the mental faculties; and although there does exist such a belief in the minds of men whose conscience is in a normal condition, still there are temptations to unbelief, and these have led men to atheism. I cannot think of an atheist unless I associate in my thoughts the words:

> "The ruling passion, be it what it may —
> The ruling passion conquers reason still."

The atheist has decided not to believe in the existence of a God, unless he can see Him and understand Him; in other words, the finite would comprehend the infinite. Following the logical method of reasoning of an atheist, the simple fact of seeing God in no way ought to prove his existence. For when you say you see a person, and that you have not the least doubt about it, I answer, that what you are really conscious of is an affection of your retina. And if you urge that you can check your sight of the person by touching him, I would answer, that you are equally transgressing the limits of fact; for what you are really conscious of is, not that he is there, but that the nerves of your hand have undergone a change. All you hear and see and touch [Pg 137] and taste and smell are mere variations of your own condition, beyond which, even to the extent of a hair's-breadth, you cannot go. That anything answering to your impression exists outside of yourself is not a *fact*, but an *inference*, to which all validity would be denied by an idealist like Berkeley, or by a skeptic like Hume. [66]

Thomas Cooper [67] said:

> "I do not say — there is no God;
> But this I say — I know not."

Mr. Bradlaugh says: "The atheist does not say, 'There is no God'; but he says, I know not what you mean by God; I am without idea of God; the word 'God' is to me a sound conveying no clear or distinct affirmation. I do not deny God, because I cannot deny that of which I have no conception, and the conception of which, by its affirmer, is so imperfect that he is unable to define it to me."

Austin Holyoake [68] says: "The only way of proving the fallacy of atheism is by *proving* the existence of a God."

If it is logical proof that is wanted, there is plenty. The following arguments, although not all meeting my approbation, are still of interest:

The *Ontological Argument* has been presented in different forms. 1. Anselm, [69] Archbishop of Canterbury (1093-1109), states this argument thus: We have an idea of an infinitely perfect being. But real existence is an element of infinite perfection. Therefore an infinitely perfect being exists; otherwise the infinitely perfect, as we conceive it, would lack an essential element of perfection.

2. Descartes [70] (1596-1650) states the argument thus: The [Pg 138] idea of an infinitely perfect being which we possess could not have originated in a finite source, and therefore must have been communicated by an infinitely perfect being.

3. Dr. Samuel Clark [71] (1705) argues that time and space are infinite and necessarily existent, but they are not substances. Therefore there must exist an eternal and infinite substance of which they are properties.

4. Cousin [72] maintained that the idea of the finite implies the idea of the infinite as inevitably as the idea of the "me" implies that of the "not me."

The *Cosmological Argument* may be stated thus: "Every new thing and every change in a previously existing thing must have a cause sufficient and pre-existing. The universe consists of a series of changes. Therefore the universe must have a cause exterior and anterior to itself.

Unity of Matter and Force.—"For if matter were not force, and immediately known as force, it could not be known at all—could not be rationally inferred."

Unity of the Life Substance in all Organic and Animal Bodies.—"A unity of power or faculty, a unity of form, and a unity of substantial composition."

[Pg 141] Unity of Animate and Inanimate Nature in Matter, Form, and Force.

Unity of the Laws of Development.—Hence we can proclaim the unity of all nature and of her laws of development.

In the beautiful words of Giordano Bruno: "A spirit exists in all things, and no body is so small but contains a part of the divine substance within itself, by which it is animated." Hence we arrive at the sublime idea, since we can in no other way account for the ultimate cause of anything, that it is God's spirit which pervades and sustains all nature. By this admission we are not led to say: "There is no God but force;" but rather, "There is no force but God." God is infinite, and therefore includes nature; but is nature all? It is all that our finite minds can discover, 'tis true; but can there not exist another nature or world unknown to us; and if so, since God is infinite, he will include that world also. Let us look to this and see what science can answer.

It will be necessary for us to consider before proceeding, what is meant by the term soul; and this becomes a somewhat difficult task, as the term has been variously applied to signify the principle of life in an organic body, or the first and most undeveloped stages of individualized spiritual being, or finally, all stages of spiritual individuality, incorporeal as well as corporeal. [76] The popular belief is, that the soul is not material but substantial, a divine gift to the highest alone of God's creatures; but scientific men, such as Carl Vogt, Moleschott, Büchner, Schmidt, Haeckel, consider the phenomena of the soul to be functions of the brain and nerves. Schmidt says: "The soul of the new-born infant is, in its manifestations, in no way different from that of the young animal. These are the functions of the infantine nervous system, with this they grow and are developed together with speech."

[Pg 142] The idea of the immortality of the soul was not aboriginal with mankind, as Sir J. Lubbock has shown that the barbarous races possess no clear belief of this kind, and Rajah Brook, at a missionary meeting in Liverpool, told his hearers there that the Dyaks, a people with whom he was connected, had no knowledge of God, of a soul, or of any future state.

Darwin remarks, that "man may be excused for feeling some pride at having risen, though not through his own exertions, to the very summit of the organic scale; and the fact of his having thus risen, instead of having been aboriginally placed there, may give hope for a still higher destiny in the distant future."

The belief in a future life amongst the civilized race of mankind is almost universally prevalent. The proofs of immortality are various. The desire that man has to live forever and his horror of annihilation is one; the good suffer in this world and the wicked triumph — this would indicate the necessity of future retribution. The infinite perfectibility of the human mind never reaches its full capacity in this life; the faculty of insight which sees in an individual all its past history at a glance is the immortal attribute and is continually on the increase; and it is possible that Aristotle was right so far as he stated that the lower faculties of the soul, such as sensation, imagination, feeling, memory, etc., are perishable. No matter if this be so or not, it is certain that in the next life, where all is perfection, only the fittest attributes will exist, the others would have perished. The doctrine of the immortality of the soul has been defended by Marhemeke, Blasche, Weisse, Hinnichs, Fecham, J. H. Fichte, and others.

Let us look for a moment at the visible universe and see if it is not reasonable, on a scientific basis, to admit of the existence of another universe, although it remains unseen to us. One can [Pg 143] not help but be struck with the fact that energy is being dissipated in this visible universe, that the visible universe is apparently very wasteful. Look at the sun which pours her vast store of high-class energy into space, at the rate of 185,000 miles per second. What will be the result of this? The answer is simple: The inevitable destruction of the visible universe. Yes, just as the visible universe had its beginning it will have its end. But there existed a power before the

"Millions of spiritual creatures walk the earth
Unseen, both when we wake and when we sleep."

If there is a life very much different from and very much higher than our present one, it is not strange we are ignorant of it. It is impossible to make a person understand anything which is entirely unlike all that has ever been seen or heard, for every idea [Pg 146] in the world that man has came to him by nature. Man [79] cannot conceive of anything the hint of which has not been received from his surroundings. He can imagine an animal with the hoof of a bison, with the pouch of a kangaroo, with the wings of an eagle, with the beak of a bird, and with the tail of a lion; and yet every point of this monster he borrowed from nature. Everything he can think of, everything he can dream of, is borrowed from his surroundings — everything. "So, if an angel should come and tell of another life, it would mean nothing to us, unless we could translate it into terms of our own experience. We could not understand a 'light that never was on land or sea.' Our ignorance is not even then a probability against our belief." [80]

As has already been stated, the visible universe must have its doom, must end as it began, by consisting of a single mass of matter; but is there not a more primitive state of matter than the matter such as we know it? Yes; and the so-called ether is that matter. It is unlike any of the forms of matter which we can weigh and measure. It is in some respects like unto a fluid, and in some respects like unto a solid. It is both hard and elastic to an almost inconceivable degree. "It fills all material bodies like a sea in which the atoms of the material bodies are as islands, and it occupies the whole of what we call empty space. It is so sensitive that a disturbance in any part of it causes a 'tremor which is felt on the surface of countless worlds.' It exerts frictions; and although the friction is infinitely small, yet as it has an almost infinite time to work in, it will diminish the momentum of the planets, and diminish their ability to maintain their distance from the sun, the consequence of which will be the planets will fall into the sun, and the solar system will end where it begun." [81]

[Pg 147] According to Sir William Thompson, the ultimate atoms of matter are vortex rings, which Professor Clifford describes as being more closely packed together (finer grained) in ether than in matter. And he says, "whatever may turn out to be the ultimate nature of the ether and of molecules, we know that to some extent at least they obey the same dynamic laws, and that they act on one another in accordance with these laws. Until therefore it is absolutely disproved, it must remain the simplest and most probable assumption that they are finally made of the same stuff, that the material molecule is in some kind of knot or coagulation of ether." [82]

The molecule of matter such as we know, then, may have been, and very probably was, produced by evolution from the atoms or vortex rings of ether, according to the theory advanced by the authors of the work called the "Unseen Universe," which I have referred to. The world of ether is to be regarded in some sort the obverse complement of the world of sensible matter, so that whatever energy is dissipated in the one is by the same act accumulated in the other; or, as Fiske describes it, "it is like the negative plate in photography, where light answers to shadow and shadow to light." Every act of consciousness is accompanied by molecular displacements in the brain, and these of course are responded to by movements in the ethereal world. Views of this kind were long ago entertained by Babbage, and they have since recommended themselves to other men of science, and amongst others to Jevon, who says: "Mr. Babbage has pointed out that if we had power to follow and detect the manifest effects of any disturbance, each particle of existing matter must be a register of all that has happened. * * * The air itself is one vast library on whose pages are forever written all that man has ever said or whispered. There in their mutable but unerring charac [Pg 148] ters, mixed with, the earliest as well as the latest sighs of mortality, stand forever recorded vows unredeemed, promises unfulfilled, perpetuating in the united movements of each particle the testimony of man's changeful will." [83]

So thought affects the substance of the present visible universe; it produces a material organ of memory. "But the motions which accompany thought," say the authors, [84] "will also affect the invisible order of things," and thus it follows that "thought conceived to affect the matter of another universe, simultaneously with this, may explain a future state." [85]

Death, then, is for the individual but a transfer from one physical state of existence to another, according to the "authors'" [86] idea; and so, on the largest scale, the death or final loss of energy by the whole visible universe has its counterpart in the acquirement of a maximum of life, the correlative unseen world. According to this theory, therefore, as the psychical or spiritual phenomena of the visible world only begins to be manifested with some complex aggregate of material phenomena, therefore it is necessary for the continuance of mind in a future state to have some sort of material vehicle also, which the ether is supposed to supply. "The essential weakness of such a theory as this," says Fiske, "lies in the fact that it is thoroughly materialistic in character. We have reason for thinking it probable that ether and ordinary matter are alike composed of vortex rings in a quasi-frictionless fluid; but whatever be the fate of this subtle hypothesis, we may be sure that no theory will ever be entertained in which analysis of ether shall require different symbols from that of ordinary matter. In our authors' theory, therefore, the putting on of immortality is in nowise the passage from a material to a [Pg 149] spiritual state. It is the passage of one kind of materially conditioned state to another." This theory, dealing with matter, should receive support by actual experience, as matter is a subject of investigation. To accept it, therefore, as being possible without any positive evidence for its support, it remains but a weak speculation, no matter how ingenious it may seem.

To support an after life, which is not materially conditioned, I agree with Mr. Fiske, that although it will be unsupported by any item of experience whatever, it may nevertheless be an impregnable assertion.

If all were to agree, what we call matter is really force, as it certainly is, for if matter were not force it would be unthinkable, being force it becomes thinkable; this point I have touched on before, but it may be well to elaborate on it a little just here. The great lesson that Berkeley taught mankind was that what we call material phenomena are really the products of consciousness co-operating with some unknown power (not material) existing beyond consciousness. "We do very well to speak of matter," says Fiske, "in common parlance, but all that the word really means is a group of qualities which have no existence apart from our minds." The ablest modern thinkers, then, believe that the only real things that exist are the mind and God, and that the universe is only the infinitely varied manifestation of God in the human conscience. It is evident, then, that *matter*, the only thing the materialist concedes real existence, is simply an orderly phantasmagoria; and God and soul, which materialists regard as mere fictions of the imagination, are the only conceptions that answer to real existence. [87]

For instance, let us see what it is we know about a table. You say you can see it; I can respond that all you are conscious of is that the nerves of your eye have undergone a change. You [Pg 150] say, I can check my sight of it by touching it; to this I reply, all that you are really conscious of is a sensation, and that something outside of you has produced it. But that all that is outside of me is anything more than the manifestation to me of a power or of God, is an inference and cannot be proven. To constant manifestations of this power, always assuming the same form and characters which can be studied, different names have been given; but that the dust of the street or beat of our heart is anything else but that peculiar manifestation of the infinite God, cannot be contradicted.

Mr. Savage says, "The movement of electricity along a telegraph-line is accompanied by certain molecular changes in the wire itself; but the wire is not electricity, neither does it produce it. Thus modern science has found it utterly impossible to explain mind either as a part or a product of matter. It is perfectly reasonable, then, for any man to believe in a purely intellectual and spiritual existence, apart from any material form or substance."

To comprehend the immortal life is an impossibility; it transcends any earthly experience of man. The caterpillar probably knows nothing about any life higher than that of his toilsome crawling on the ground; but that is no proof against the fact that we know he is to become a butterfly. The boy knows nothing about manhood, and cannot know. Though he sees men and their labors all about him, he has and can have no conception whatever of what it means to be a man; it transcends all experience. [88] "The existence," says Fiske, "of a single soul, or congeries of psychical phenomena, unaccompanied by a material body, would be evidence sufficient to demonstrate this hypothesis. But in the nature of things, even were there a million such souls round about us, we could not become aware [Pg 151] of the existence of one of them; for we have no organ or faculty for the perception of soul apart from the material structure and activities in which it has been manifested throughout the whole course of our experience. Even our own self-consciousness involves the consciousness of ourselves as partly material bodies. These considerations show that our hypothesis is very different from the ordinary hypothesis with which science deals. *The entire absence of testimony does not raise a negative presumption, except in cases where testimony is accessible."*

My object has not been to prove the purely spiritual theory of a future life, but to show that it is a theory that intelligent people can entertain as a foundation for their belief "in the hope of immortality." But that the spiritual life instead of the material life is the state in which we can hope for immortality, I think there can be no question; and such was the opinion of Paul [89] when he wrote: "Now this I say, brethren, that flesh and blood cannot inherit the kingdom of God, neither does corruption inherit incorruption.... So when this corruptible shall have put on incorruption, and this mortal shall have put on immortality, then shall be brought to pass the saying that is written, 'Death is swallowed up in victory.'

> O death, where is thy sting?
> O grave, where is thy victory?"

Footnotes:

[1] The Law of Disease, in College Courant, Vol. XIV.

[2] Winchell. Evolution, p. 113.

[3] Comparative Zoology, p. 43. 1876.

[4] Huxley. Physical Basis of Life.

[5] Johnson, Ency.

[6] Comparative Anatomy — Orton, p. 32.

[7] Analytical Anatomy and Phys.—Cutter, p. 16.

[8] Biography of a Plant.

[9] See Huxley—Invertebrate Animals, Anatomy of.

[10] Phys. Basis of Life.

[11] Beginnings of Life, p. 104, Vol. I.

[12] Monthly Micros. Jour., May 1, '69, p. 294.

[13] Chem. and Phys. Balance of Organic Nature, 1848, p. 48 (trans.).

[14] Inaugural Address, Aug. 19, 1874.

[15] Haeckel—Hist. of Creation.

[16] See Haeckel—Evol. of Man.

[17] Evolution of Man, Vol. II, p. 445.

[18] Johnson's Cyclopedia, Article "Evolution."

[19] Sumner, in Johnson's Cyc.

[20] Christian Union, Vol. XIII, No. 17, p. 322.

[21] Gen. i. 1.

[22] St. John i. 1.

[23] St. John i. 3.

[24] Hist. of Creation, p. 8.

[25] *Ibid.*, p. 324.

[26] Heb. xi. 3. Revised English Ed.

[27] *Loc. cit.*, Vol. I, p. 323.

[28] *Loc. cit.*, Vol. I, p. 324.

[29] Indications of the Creator.

[30] Evolution and Progress, p. 26, Rev. Wm. I. Gill.

[31] Natürl. Schöpfungsgesch., pp. 643-5.

[32] Paget, Lectures on Surgical Pathology, 1853, Vol. I, p. 71.

[33] Ueber die Richtung der Haare am menschlichen Körper.

[34] Pop. Sci. Monthly, June, 1879, p. 250.

[35] See Sci. Am., May 18, 1878.

[36] Source of Muscular Power, Proc. Roy. Inst., June 8, 1866. Am. I. Sci., II, xlii, 393, Nov. 1866.

[37] Comparative Zoology, p. 45.

[38] Correlation of the Vital and Physical Forces, p. 54.

[39] On the time required for the transmission of volition and sensation through the nerves, Proc. Roy. Inst.

[40] Comparative Zoology, p. 165.

[41] Sci. Amer., Nov. 13, 1876, p. 328.

[42] Marshall, Outline of Physiology. Amer. Ed., 1868, p. 227.

[43] Macmillon's Magazine, Pop. Sci. Monthly, April, 1876.

[44] "Principles of Psychology," 1869, No. 20, p. 24.

[45] J. S. Lombard, N. Y. Med. Jour., Vol. V, 198, June, 1867.

[46] *Loc. cit.*, p. 23.

[47] The apparatus employed is illustrated and fully described in Brown-Sequard's Archives de Phys., Vol. I, 498, June, 1868. By it the 1-4000th of a degree Centigrade may be indicated.

[48] L. H. Wood, "On the influence of mental activity on the excretion of phosphoric acid by the kidneys." Proc. Conn. Med. Soc., Nov., 1869, p. 197.

[49] *Loc. cit.*, p. 24.

[50] Address of Dr. F. A. P. Barnard, as retiring president, before the Am. Ass. for Adv. of Sci., Chicago meeting, Aug. 1868. "Thought cannot be a physical force, because thought admits of no measure."

[51] Derivation hypothesis of life and species, forming fortieth chapter of his Anatomy of Vertebrates, republished in Am. Jour. Sci., II, xlvii, 33, Jan. 1869.

[52] Prehistoric Times, p. 354, by Lubbock.

[53] Madness in Animals, Jour. Mental Sci., July, 1871. Dr. W. L. Lindsay.

[54] Facultés Mentales des Animaux, 1872, Tom. XI, p. 181.

[55] Primeval Man, 1869, pp. 145-147.

[56] Prehistoric Times, 1865, p. 473.

[57] "Conferences ser les Théorie Darwinienne," 1869, p. 132.

[58] Philosoph. Trans., 1773, p. 262.

[59] Prof. Whitney, p. 309.

[60] Phys. and Pathol. of Mind. Dr. Maudsley. 3d ed., 1868, p. 199.

[61] Nature, January 6, 1870, p. 257.

[62] Problems i. 21.

[63] Johnson's Cyc. Article "Faith." C. P. Krauth.

[64] Darwin's Descent of Man, p. 117.

[65] See Descent of Man, p. 96.

[66] See Tyndall's Belfast Address.

[67] Purgatory of Suicides.

[68] Thoughts on Atheism, p. 4.

[69] Monologium and Proslogium.

[70] Meditations de Primaphilosophia Prop. 2, p. 89.

[71] Demonstration of the Being and Attributes of God.

[72] Elements of Psychology.

[73] Thoughts on Atheism, by Holyoake, p. 4.

[74] Proverbs xvii. 22.

[75] Henry Ward Beecher.

[76] See W. T. Harris. Johnson's Encyc. "Soul."

[77] Unseen Universe.

[78] Rood. "Sound," Johnson's Encyc.

[79] See R. G. Ingersoll's Lecture on Hell.

[80] Savage.

[81] "The Unseen World." John Fiske, p. 21.

[82] Fortnightly Review, June 1875, p. 784.

[83] Ninth Bridgewater Treatise.

[84] Of the Unseen Universe.

[85] Anagram. Nature, Oct. 15, 1874.

[86] Of the Unseen Universe.

[87] Fiske. Unseen World, p. 52.

[88] Savage. Relig. of Evol., p. 246.

[89] 1 Corinthians, xv., verses 50-54 (Part of). *Revised English Ed.*, 1877.

www.ingramcontent.com/pod-product-compliance
Lightning Source LLC
Chambersburg PA
CBHW031424210526
45464CB00005B/2041